MW00630058

When Science Meets Power

WHEN SCIENCE MEETS POWER

Geoff Mulgan

polity

First published in 2024 by Polity Press

Polity Press
65 Bridge Street
Cambridge CB2 1UR, UK

Polity Press
111 River Street
Hoboken, NJ 07030, USA

ISBN-13: 978-1-5095-5306-8

A catalogue record for this book is available from the British Library.

Library of Congress Control Number: 2023937258

Typeset in 11.5 on 14 pt Adobe Garamond
by Cheshire Typesetting Ltd, Cuddington, Cheshire
Printed and bound in Great Britain by CPI Group (UK) Ltd, Croydon

Contents

Detailed Contents

Acknowledgements

I've benefited from the insights of very many thinkers and doers in writing this book – too many to mention all of them – and from a career that has taken me between the worlds of government and science, universities and practical action. I may have accumulated some prejudices along the way, but I hope the breadth of my background has helped me to see patterns that weren't so obvious to others within a single field.

While writing the book I tried some of the ideas and arguments out with many colleagues, both in governments and in science, including through my role at UCL Science, Technology and Public Policy (STEaPP); through engagement with the Geneva Science Diplomacy Accelerator (GESDA), the OECD's science and technology team; government science advisers such as Sir Peter Gluckman, Maarja Kuusma and Joe Biden's science adviser Eric Lander; Tatjana Buklijas and Kristiann Allen at the International Network of Government Science Advisers; Anja Kaspersen (ex UN now at IEEE); Effy Vayena at ETH in Zurich; Helen Pearson at Nature, who commissioned a piece on synthesis which prompted part of the book; the team at Science Advice for Policy by European Academies (SAPEA); David Mair and others at the European Commission Joint Research Centre; many colleagues at the OECD, UNESCO and UNDP (including through the STRINGs programme on steering research for the global goals, which fed into the chapters on international options); James Wilsdon and the 'Research on Research Institute' programme; Robert Doubleday and others at the Centre for Science and Policy (CSAP), in Cambridge. I benefited from useful inputs from colleagues, including Professors Jon Agar, Arthur Peterson, J.C. Mauduit, Jo Chataway and Chris Tyler, and PhD students Basil Mahfouz and Alex Klein. I am also particularly grateful to Jonathan Skerrett at Polity for commissioning the book and for very useful feedback. And, naturally, I have drawn on the writings of many others whom I have never met, and gained from working with others who are no longer alive, including the sometimes rivalrous Bruno Latour and Pierre Bourdieu, and the UK science adviser Sir Robert May.

None of those mentioned above are, of course, in any way responsible for any errors of fact or judgement in this book and, though I hope they would all agree with its main argument, I doubt that any will agree with all of it. My hope is that it will promote the kind of debate – and healthy friction – which lies at the heart of the best science and the best politics.

Introduction
The science–politics paradox

Many of the world's most prosperous cities – including London, Hamburg and Rome – have within them a research centre studying what are called 'BSL4' organisms. These are dangerous organisms, which pose a 'high risk of life-threatening disease, aerosol-transmitted lab infections, or related agents with unknown risk of transmission'. The Labs that study them are also labelled as BSL4. They are familiar from the evening news as places full of people wearing what look like space suits, and they often show up in Hollywood films.

It's estimated that there are some 69 of these either in operation or construction worldwide. Most are located in urban areas. Most of the people living nearby, which probably includes many readers of this sentence, are unaware of them. Most might be shocked if they knew what was being done within them, often with only weak oversight and regulation,[1] and would wonder why they are located in highly populated cities, like the one in Wuhan which some thought might lie behind the COVID-19 outbreak.[2]

Much of their work is necessary. But they are making judgements about risk that have huge significance for everyone else, for example when they experiment with combining viruses, or amplifying their harm or transmissibility. Yet there are no global agreements on how they should be run, no official registries of where they are, what they do, or how safe they are.

This is one of many examples of a gulf that has grown up between science, which is often necessary and inspiring but also often opaque and secretive, and public interest or public dialogue, or what some would call common sense. They are examples of the challenge every society has of exercising power over knowledge and reminders that while science may save us, whether from diseases and natural threats like asteroids, it may also kill us, whether with a nuclear Armageddon, invented pathogens, or wayward intelligence.

The ambiguities of science and technology are evident in the impacts of the many technologies that have seen exponential improvements in recent years. The processing power of computers has roughly doubled every two years in line with Moore's law. The cost of sequencing the full human genome fell from around $100 million two decades ago to $100 today, while the cost of solar powered electricity fell almost 90 per cent in the 2010s alone.[3]

These advances happened alongside stagnant incomes for many in countries such as the US, stagnant wellbeing, declines in social connectedness and mental health, as well as worsening global ecological indicators and signs of potential systems collapse.

Young people reflect these paradoxes: most are very positive about science and technology, more than their parents' generation, and technology plays a big part in their lives. According to one survey of 20,000 young people across the world, 84 per cent of them say that technical advancements make them hopeful for the future. But this enthusiasm for science combines with pessimism. In sixteen out of twenty countries, more young people believed the world was becoming a worse place to live than believed it was becoming better, with many of the reasons cited being indirect effects of science, from workplace automation and climate change to the harmful effects of social media.

Science is the most extraordinary collective achievement of the human species – a set of methods, mindsets, theories and discoveries that have changed every part of our lives. But these paradoxical patterns show that a powerful method for amplifying human intelligence is not always so intelligently directed.

So you might expect that the question of how to govern science, how to mobilize its benefits but avoid its risks, not just in biohazard labs but also in everything from artificial intelligence and food systems to space warfare, would be one of the most important questions of our times.

But how? Scientists have long argued that they should be given the maximum freedom to explore and discover. While there are good arguments for this, particularly in more fundamental science, it becomes ever less plausible the closer science and technology come to daily life. Although scientists are typically intelligent, thoughtful and decent people, it's not obvious that they can be trusted to govern science, any more than the military can be put in charge of wars. Science alone can be tunnel-visioned: it needs other perspectives to show how to avoid harm.

This is where institutions come in. Our societies are made predictable and manageable through institutions. It's through institutions that a public interest comes to be refracted. Within nations, an array of funders, regulators, agencies, commissions and parliamentary groups try to steer science and innovation. But there are glaring governance gaps – gaps where institutions are needed, whether in relation to AI and cybersecurity or synthetic biology. Moreover, public influence over science has declined in recent years as the proportion of R&D dominated by big firms has grown. Amazon, for example, spent $40bn on R&D in 2020, more than all but a handful of countries (the UK's public R&D budget that year was around $14bn, Finland's around $2.3bn).[4]

These gaps are even more evident at a global level where there is little effective governance. We have a World Bank, an International Monetary Fund and numerous other funds and development agencies. But the world lacks institutions charged with reflecting on science, its aims and its methods, or judging whether the world's research capacity is directed to the right tasks and with the right methods. Scientists have achieved many breakthroughs in global cooperation, often below the radar of formal politics – from the rules of the Internet to the management of Antarctica and nuclear non-proliferation. Yet the best available data show a striking lack of alignment between what the world has decided are its top priorities – summarized in the Sustainable Development Goals – and the priorities of science and technology.[5]

Scientists are periodically involved in public calls to guide, constrain or rein in powerful technologies, particularly artificial intelligence. But, as I show later, these are usually so vague, and so devoid of any plan of action or any language for thinking about governance, that they have little impact.

In the nineteenth century, as constitutional monarchy became the norm in much of Europe, it was said that monarchs now reigned, but didn't govern. Science is now in an opposite position. It governs but doesn't reign and is only loosely accountable for the power it exercises.

Some of the reasons for this lie in the blind spots and biases of science itself. Scientists say they can't make policy, that they have neither the skills nor the inclination to do so, and they fear being tarnished if they get too close to the grubby, compromised world of actual government. Yet their *de facto* power makes this position of detachment increasingly implausible.

Politics should be the answer, since it is the main way we make collective decisions. But politics looks ill-suited to the task of governing science. The dominant forms of modern politics were shaped in the nineteenth century: rule by representatives concentrated in parliaments in capital cities, with periodic elections, manifestos and programmes. There has been relatively little advance since then, despite many experiments on the periphery (from citizen assemblies and deliberations to virtual parliaments). Instead, politics often looks petty, short-term, half-informed, or irrelevant. The ways in which politicians are recruited and promoted don't fit well with the tasks they have to fulfil and their roles are almost unique in being so unsupported by professional training – most learn on the job.

I've sometimes played a slightly mean trick on senior figures in politics.[6] I ask them if they could give a five-minute talk on how the Internet works. Almost none can, though they use it for many hours a day. They know next to nothing about how it functions, or about the material reality of undersea cables and switches or the organization of addresses and protocols.[7] Essentially for them it is magic, which perhaps helps explain why governments and parliaments found it so hard to respond intelligently as the Internet transformed so many areas of life, both for better and for worse.

So, for politics to play the roles that only politics can play we need a radically reformed politics. This is what I call the 'science–politics paradox': only politics can govern and guide science in the public interest, but politics has to change to be able to do this: to become more knowledgeable, more systematic in its methods, and some of the time, more scientific, benefiting from what I call the 'new curriculum for power' that encompasses data and systems, complexity and psychology as well as politicians' more traditional grounding in law and economics.

I use Hegel's story of the master and the servant as a way to make sense of this dynamic. Politics, the putative master, has nurtured a servant who now greatly outstrips the master in terms of capability and knowledge. Science has gained a *de facto* sovereignty of its own, that sits along the traditional sovereignty of politics: the servant has to some extent become a master. Most of our collective decisions now involve science – from pandemics to climate change, the upbringing of children to clean air – and that collective knowledge now makes a claim that complements the claims of votes, or the desires of citizens.

But science has little to say on how we make these collective judgements, and little to say about questions of meaning, or wisdom. It can tell us what is, and what might happen, but it can't tell us what matters, or what we should care about. For that we need politics, in its broadest sense. The idea that governments can simply 'follow the science' quickly falls apart on inspection.

Here I draw on Aristotle. He distinguished between ethics, which concerns what is a good life for an individual, and politics, which is concerned with the good life for a community. He believed that the health of the polis was essential to the full realization of human potential.[8] He saw political science as a master discipline, an 'architectonic', which sits above the other disciplines, and, he wrote in the *Nicomachean Ethics* that 'since political science uses the rest of the sciences, and since, again, it legislates as to what we are to do and what we are to abstain from, the end of this science must include those of the other sciences, so that its end must be the good for humanity'.[9] Two thousand years later, politics can still override all other fields and disciplines through its power to make laws.[10]

In doing this it draws on ethics but goes broader.[11] Indeed, most of the decisions to be made about science and technology go far beyond ethical reasoning: they involve very political judgements about who benefits and loses and they are highly contextual. The fashion for creating centres around the ethics of science (from biosciences to AI) is an understandable response to the failures of politics, and often produces intelligent commentary. But I suspect in the future it will be seen as a category error. Ethics alone cannot tell us how to design a system of welfare, when to fight wars, how to tax or how to police, or how to guide powerful new fields of science.

It is a political question, not an ethical one, whether power needs to be mobilized to block, accelerate or guide science and technology, with the authority to inspect, analyse and assert rules and laws. It is a political question whether power needs to be mobilized to influence how technology is itself enabling new forms of power, such as monopoly, predation or abuse. And it is a political question whether power needs to be mobilized to distribute the benefits of new knowledge (such as genetic enhancement).

During the course of the book, I dive into the nature of these relationships with power. I describe the complex history of state involvement in science; how states saw science as the means to military prowess or economic prosperity; the rising concern with risk; and the practical problems faced by governments and parliaments grappling with science advice. I describe the clashing logics of science, politics and bureaucracy and the ways in which these logics have a life of their own.

My conclusion is that we need a simultaneous scientization of politics and a politicization of science, reinventing both, so as to cultivate sciences that are reflexive and self-aware of their own limits and politics that are sufficiently well informed to guide processes that are often opaque, uncertain and hard to grasp.

This is not an argument for scientists becoming partisan. Quite the opposite. The more scientists appear *parti-pris*, using their authority as scientists to endorse views that have nothing to do with science, the less they will be trusted.[12] The more they appear closed and narrow in their thinking, the less reason we will all have to take their conclusions seriously. For me, the politicization of science is more about scientists taking responsibility for the state of their society and being willing to engage in argument and debate about its priorities. It is about acknowledging that many of the most important decisions to be made about science are essentially political.

The heart of my case is an argument for metacognition. Metacognition is the crucial skill schools teach children: a skill of thinking about how to think, and knowing what the appropriate ways of thinking are for different tasks.[13] In science and technology there are often very strong systems for cognition but only weak ones for metacognition, for reflecting in a rounded way on complex choices. This kind of metacognition amplifies the spirit of science, the commitment to exploration and doubt, but it sometimes challenges the practice. It also amplifies the best of democratic politics – a willingness to engage with other people and other views – and challenges its tendency to become narrow.

That requires a skill in looped rather than linear thinking, since multidimensional knowledge – knowledge that encompasses ethics, politics and much more – ultimately has to take precedence over the less dimensional knowledge of individual scientific disciplines. To handle a complex task like managing a pandemic, averting climate change, banning rogue AIs or fighting a war requires multiple types of knowledge, of which scientific knowledge is only one, and not always the most important. It's necessary, in other words, to zoom out before zooming back in.

The mark of a mature political system, I argue, is that it has many different ways of mobilizing knowledge, suitable for tasks with varying degrees of technical and moral complexity, varying links to the daily life of citizens, and varying degrees of uncertainty, and that it can explain why different ones are used for different purposes.

Metacognition underpins synthesis. Science has extraordinarily strong tools for analysis and discovery. But it has surprisingly weak methods for thinking across boundaries or for synthesis. This became very apparent during the pandemic as some scientists became very powerful – but could not articulate how they would weigh up physical health against mental health, the needs of the economy or education. They could be hugely impressive within their domains, for example modelling the risks of transmission, or accelerating the development of vaccines, but were oddly inarticulate across domains. Yet many of our big challenges – from the complexities of shifting towards a hydrogen-based economy to population mental health – require exactly this kind of synthetic or holistic thought and action.

Science often governs itself. When it doesn't, it is still much more often directed to the interests of states or of corporations than it is to the wider public interest. This is one field that has yet to democratize its governance, to ensure that common knowledge serves common interests. As I show, there is a wide gap between what the public say they want science to focus on and where brainpower is actually directed. This may be why, in the UK for example, a majority say that R&D does not benefit them.[14]

To better serve the public we also need better anticipation. Science produces new knowledge but also new risks and so, for any society, and for the world as a whole, the ability to understand, spot, anticipate and prevent is vitally important. Yet in most fields we have only weak institutions to do this. The IPCC attempts to anticipate trends in climate change – but is interesting in part because it is such an exception, with no equivalents in fields like artificial intelligence.

Anticipation in conditions of uncertainty points to actions that need to be different in nature from traditional laws and programmes. Precisely because of the uncertainties surrounding science they have to be revisable decisions – decisions that include clarity on the triggers, or new facts, which would require a change to the decision. Regulation needs to become more 'anticipatory' – able to anticipate technological change and to shift quickly in the light of its actual patterns, whether in relation to drones or quantum computing, genomics or self-driving cars.[15] And, rather than one-off inquiries and commissions, we need more permanent, continuous assessment of benefits and risks: what I call 'science and technology assemblies', which bring together experts, politicians and the public at every level, from cities to nations to the world, and are supported by well-curated knowledge commons. These need to be both political – in the narrow sense of engaging with interests and values in the present – but also super-political, in the sense of seeking to take account of the interests of future generations.

These arguments about the 'how' of government and governance reflect the shifting nature of truth. States rest for their legitimacy on claims about truth, sometimes arbitrary and sometimes accurate. Science too aspires to the discovery of truths. Yet we are in a time when truth can seem slippery, when innumerable fakes, false images, videos and misleading claims proliferate, with deception becoming cheaper and more commonplace, making it harder than ever to know what to believe. In this context infrastructures of verification become even more important, and professions with a vocation for proof and truth become ever more socially vital. The basic methods of science – which involve scepticism and rigorous method to get closer to truths – matter not just to the work of science itself but to almost everything else. Indeed, the original motto of the Royal Society, founded in London in the seventeenth century – *nullius in verba* – is even more relevant in an age of wars over truth (the motto essentially means: don't take anyone's word; instead, test, interrogate, probe).

This – the wider value of science – makes it even more important that scientists engage. I argue for a 'relational turn': that science needs to work harder not just at explaining, but also at listening, responding, and opening up to democratic input. It is still common to hear scientists talk as if communication was enough to ensure trust. This is wrong.

Scientists will be respected for their expertise but, in the long run, they will only be fully trusted if they are seen to care about the interests of the public.

The core insight of politics is that it is only through expression, argument and competition that we discover and express common interests. The core insight of science is that it is only through detached observation, experiment and scepticism that we discover useful truths. And the core insight of bureaucracy is that it is only through creating institutions, roles and rules that we make things happen.

This book makes the case for fusing these insights into a new generation of institutions to shape science, supported by new logics, that can help whole societies think together about their choices and implications, from data and evidence to imaginative speculation. Their task is to think and act synthetically, connecting the four stages that are vital for any governance of science and technology: analysis and observation; assessment and interpretation; action using the full range of possible tools, from laws and regulations to funds; and then adaptation in the light of what happens.

In emphasizing action and learning I take a different view from that of many writers on science, particularly in the field of science and technology studies, some of whom have opted for detachment: observing and analysing but steering clear of prescription. I suggest that their stance is a symptom of a more widespread 'dynaphobia': a fear that any engagement with power will be corrupting (fear that mirrors the excessive 'dynaphilia' of many politicians and officials, who perhaps love power too much and knowledge too little).

Many academics avoid making proposals, designing options and advocating for them, preferring to stay in the safer space of observation and critique. The result is a deficit of designs that becomes very obvious when, for example, societies need better ways of governing technologies such as artificial intelligence. At a time when we badly need useful options for the synthesis of science, politics and ethics, whether in government or business, many of those with the deepest knowledge of the issues are mute.

The job of creating and implementing such designs is a truly political task. With any emerging field of science and technology, a society has to decide whether to encourage it or discourage it; to fund or defund it; to establish new rules or institutions to guide it or to opt for benign neglect.

These decisions are partly fractal, made not just in laws but also in the conscience of individual scientists, the managements of firms or foundations, shaped by media and movements. But many of the most important decisions at some point return to politics.

PART I

How Science Meets Power

1

Uneasy interdependence

Science is all around us and, if we look into the future, its significance is only set to grow. Science shapes our health and fuels new technologies increasingly integrated into our bodies, our homes and our cities. It illuminates the cosmic context of our lives and reveals the minutest details of life. It is a source of wonder, inspiration and awe.

Now, unlike our predecessors, we often can't help but see things through a scientific lens. A patch of parched grass may be seen as the result of climate change. Unruly children may be interpreted through the lens of the science of parenting. A fast-food shop may be looked at through what we know of nutrition or obesity. In all these ways scientific reasoning connects to, and sometimes displaces, other lenses: land seen primarily through the lens of belonging; food through the lens of pleasure and gratification; children through the lens of belief or the sanctity of the family.

But, to the extent that science has grown, so too has it become more dangerous. Some still hold to a view of science as cool, calm and ordered, with the quiet hum of laboratories and people in white coats bringing sanity and rationality where once there was chaos, capriciousness and violence.

However, this picture is not accurate. New knowledge is unsettling and destabilizing. It answers some questions but generates new ones. Ever more of the risks we face are the results, either direct or indirect, of scientific progress, from the development of a carbon-based industrial civilization to nuclear and biological weapons, rampant artificial intelligence to genetically modified organisms. New knowledge destroys old jobs as well as old sources of authority. It reveals new areas of ignorance and creates new anxieties and justified fears. This is the paradox of science that it is both ever more vital and ever more dangerous.

This is obvious when we look at the countless threats that originate in science, from pathogens to pollutants. The Commission for the Human Future, in 2022, for example, highlighted ten potentially catastrophic threats to human survival, a list similar to many other ones. These are (in no particular order):

1. **Decline** of natural resources, particularly water.
2. **Collapse** of ecosystems and loss of biodiversity.
3. **Human population growth** beyond Earth's carrying capacity.
4. **Global warming** and human-induced climate change.
5. **Chemical pollution** of the Earth system, including the atmosphere and oceans.
6. Rising **food insecurity** and failing nutritional quality.
7. **Nuclear weapons** and other weapons of mass destruction.
8. **Pandemics** of new and untreatable disease.
9. The advent of **powerful, uncontrolled new technology**.
10. National and global **failure to understand and act preventatively** on these risks.

Most are the direct or indirect results of a science and technology based civilization, and all are amplified by the last risk listed here. These together make the case for new arrangements and institutions on the cusp of science and politics to better avoid 'national and global failure'. But how? And how can they collaborate?

Both science and politics have very long roots but have taken distinctive forms in the modern world: politics in the form of states, parties, parliaments and programmes, science in the form of disciplines, labs, methods for experiments or peer review. Both promote knowledge and effective action through a mix of competition and cooperation. Both rely heavily on words and prose. Both use a surprisingly similar structure to link thought and action. They observe – what's happening and what matters. They interpret. And they act.

Yet there is a fundamental difference in how they think. Politics is infinitely flexible. There is no such thing as a 'political truth'. All that matters is what works, for now, often with a very short time horizon. Science by contrast is dogmatic in the sense that it has a rigid view of what methods are acceptable and which are not (though its dogma is that there are no dogmas – everything is open to questioning). It polices its frontiers to root out heresies and falsehoods. It can take the long view, and its methods of analysis are deep, but also linear.

The mentality of science is by its nature sceptical and cold. Indeed, this is its greatest strength. Faced with any claim, it asks us to question, prod, doubt and interrogate. This is what distinguishes it from myth or narrative. It helps us to get closer to truths of all kinds. But it gives us little comfort. The mentality of politics is very different. It seeks to reassure and make sense, while representing, channelling and reflecting our collective needs and desires, rooted in time and space and in the lives we live. Science has relatively little to say about what matters, though it can warn and encourage. Politics has relatively little to say about facts, though it needs facts to guide its diagnoses and prescriptions.

Both are radically incomplete. Science achieves most of its impact in the world through the addition of engineering, which has a very different logic and way of working (though increasingly science and engineering are interwoven, for example in the frontiers of artificial intelligence). It has to combine with other kinds of reasoning – ethical, political, pragmatic – to make any judgements about what to do: science alone cannot tell us what counts as gender in sports or whether nuclear power is a good answer to climate change. It is a vital input but, in order to be useful for action, it has to accept its place alongside other types of knowledge that are just as relevant.

Politics too is incomplete. It should be a synthetic practice, drawing on many kinds of knowledge and aware of its own deficiencies and blind spots. But often it narrows down to a caricature. Its nature is to be flexible but this can become a pathology, without regard for facts, consistency or practicality. Politics that is only politics serves the public poorly. And so politics only works when it combines with other ways of knowing and doing.

Science and politics need each other to survive. Science needs the patronage of politics, politics needs the solutions of science. But they also compete for authority, for resources and for recognition. Their uneasy symbiosis casts a new light on old dilemmas. For two thousand years political philosophers have debated whether superior knowledge, or backing from other citizens, provides a sounder basis for legitimacy.[1] We experience a similar tension in our own lives – the tension between what we know and what we do, feel or identify with. It is a rare person who acts straightforwardly on the basis of knowledge, whether in diet and fitness, relationships, voting choices or career choices. Instead, we struggle with the tension between what we know and what we are.

In politics the tension between the idea of legitimacy based on expertise, and legitimacy based on the expression of civic will, is equally unavoidable. No government can be entirely 'evidence-based'; none can defer all decisions to scientists and experts; and none can ignore the moods, hopes and fears of the public.

Yet no government can be entirely driven by public desires either, since these will be incoherent and inconsistent, and no public would itself wisely choose to be driven by its own choices. Again, there is a parallel with our own individual lives: most of us put in constraints, commitments and arrangements that protect us from our own unstable volition.

In relation to politics, however, it is hard to articulate precisely what the limits of popular sovereignty should be. We prefer the myth of public wisdom: the claim that on balance, publics will tend to make the right choices, rather than the more accurate perception that there is at best a loose correlation between wishes and outcomes. This was well described by the Israeli political scientist Yaron Ezrahi as 'an unsettling empty dark space at the foundation of political order'. That dark space has become increasingly unstable as science has grown and as politics has, at times, reacted aggressively to the challenge this implies.

In the case of politicians like Donald Trump and Vladimir Putin, their advisers and ideologists, there is no embarrassment in creating their own facts, their own universes of meaning, and speaking contemptuously of science when it doesn't serve them. These are the more straightforward cases, where the world of myth hits the world of science. But just as common are much less easy cases, where facts and science, values and politics, intermingle without it being so obvious who is on the side of virtue.

An intimation of the possibly uncomfortable future relationship between science and politics could be seen on 20 July 2021 when Dr Anthony Fauci, Chief Medical Adviser to the US President, appeared before a Senate hearing. He was there to challenge the claim that the US government had funded research in Wuhan that could have led to the leak of the pathogen that caused the COVID-19 pandemic. In response to a question from one of the Senators he replied: 'I totally resent the lie you are now propagating.'[2]

Such moments, when cool science hits hot politics, have become ever more common. In this particular case, it remains unclear whether COVID-19 was indeed the result of scientific research that then leaked (though it looks more likely that it came from animals in the Wuhan market, with raccoon dogs passing it on from bats). Eminent figures could be found on both sides of the argument.[3] But what was not in doubt was that the US NIH had funded experiments. It was certain that it had financed activity in Wuhan that involved coronaviruses, with several organizations funded to do what's called 'gain-of-function research' (which attempted to increase the virulence or harms of viruses), constrained only by weak provisions that funders should be informed if there were dramatic results. And it was certain that actions taken to allay suspicions – with key players who were implicated in the problems being recruited to investigate them and key evidence suppressed – instead fuelled suspicions.

This incident – which concerned a pandemic that caused some twenty million deaths globally – remains murky.[4] But it highlighted many uncomfortable issues, including the risky patterns of some research, which pushes back the boundaries of human knowledge but can appear to lack much wisdom or common sense.

The US Congress concluded that they had lost control, during a year which should have been one of unmitigated triumph for science, having created and distributed a series of effective vaccines at extraordinary speed. Some ambitious politicians saw an opportunity: Ron DeSantis, Governor of Florida, called in late 2022 for a grand jury to investigate 'any and all wrongdoing' with respect to COVID-19 vaccines, signalling his intention to be even more sceptical of science than President Trump (only 17% of conservative Republicans now report having a lot of trust in scientists, compared to 67% of liberal Democrats).[5] Here are signs of just how much science had become simultaneously indispensable but also problematic.

That same year a Wellcome Trust poll found that 80 per cent of people from 113 countries said they trusted science either 'a lot' or 'some', a level of support that other fields can only envy. But that success, too, masked uncomfortable patterns. 44 per cent of Americans do not believe that human activity is causing climate change,[6] while in South Africa fewer than a fifth believe that it is.[7] Shortly after the pandemic I walked down my local high street (in Luton, a medium sized town in England) to see a series of stalls (some Christian, some Islamic) explaining that COVID was a punishment from God for various sins committed and proclaiming that science had failed and only religion could answer the true questions of life. Some shared the opinion of 40 per cent of Americans that we are living in 'end times'.

What mattered to them was a world away from what seemed obvious to scientists, though it was paradoxical that quite smart science was playing its own role in undermining science (it's estimated that nearly half of the Twitter accounts spreading messages on the social media platform about the pandemic are likely to have been bots).[8]

The arguments about COVID and vaccines were particularly intense examples of the evolving struggle between science and both its enemies and its sceptical friends. But they were hardly unique. During the same period the European Parliament debated new laws on artificial intelligence (I was on one of its advisory committees, part of STOA – Science and Technology Options Assessment). AI is extraordinary, impressive, part of our daily life and also part of our daily dreams and nightmares. As we will see, many of the scientists at the heart of it have tried to define ethical rules and limits to their own work, though with only limited success.

Politicians struggled to know what they should do. On the one hand they were told that AI was vital to the future prosperity of their continent, which was already slipping behind the US and China on the frontiers of computer science. On the other hand, they could see how manipulative and dangerous AI could be to their citizens. As a compromise they proposed to ban outright a range of algorithms that were deemed high risk and further pushed the principle that algorithms should be transparent and explainable. Some argued these were quite impossible to implement, as algorithms, and AI using neural nets, became ever more opaque and complex.

Yet the political pressures to regulate AI were unavoidable. In 2020 thousands had marched in London against an algorithm that had determined school exam grades. In the Netherlands an AI algorithm had determined incorrect payments for social security to thousands, causing much misery (and forcing the government to resign). During the same period – 2021 and 2022 – China introduced a clutch of new rules on

AI, establishing ethical principles and limits, including the first city-level legislation in Shenzen in late 2022 and the first provincial rules in Shanghai the same month, all trying to establish different categories of risk and constraint.

Here, too, science was everywhere but, again, also problematic, pushing issues onto the political agenda that politicians struggled to understand but which scientists also lacked the intellectual tools to judge. Look closely and this is now a normal situation not an exception. Dozens if not hundreds of science assessments are underway in any country at any time. They may be estimating the cost of nuclear waste and deciding who should be responsible – since if it's the industry, investment might stop – or trying to assess how much to hope for fusion, which always appears just thirty years from fruition, or whether to adjust the rules for gene-editing.

Some judgements are about when and how to accelerate technologies – like quantum computing – to ensure that nations play a part in coming economic booms. Others are about when and how to slow them down, as when decision-makers realize that, while quantum computing could drive new industries, it could also undermine privacy on the Internet, demolishing the cryptography on which cloud computing and messaging systems like WhatsApp depend, as well as having potentially catastrophic implications for national security.[9]

Science as threat; science as failure; science as prompt for new rules. All are now everyday dimensions of science and all are highly political in every sense of the word. In the words of Peter Gluckman, former chief science adviser to the Prime Minister of New Zealand, a host of problems now requires decisions that are simultaneously scientific and political, including 'eradication of exogenous pests [. . .], offshore oil prospecting, legalization of recreational psychotropic drugs, water quality, family violence, obesity, teenage morbidity and suicide, the ageing population, the prioritization of early childhood education, reduction of agricultural greenhouse gases, and balancing economic growth and environmental sustainability'.[10]

Yet our current arrangements to manage the boundaries of science and politics are no longer adequate. They neither acknowledge the growing sovereignty of science – its own claims to authority and legitimacy – nor its weakness, which is a lack of capacity for synthesis, integrating scientific insights with other types of knowledge and grasping what truly matters. An ever-reducing proportion of political decisions makes no use of science, and, conversely, an ever-reducing proportion of scientific decisions is untouched by politics.[11]

But how should we use and govern science? How should a society use the best available knowledge to guide it? Should we worry more about out-of-control scientists or democratic politicians with little knowledge?

There are surprisingly few good examples of a society engaging in a thoughtful, rounded way with a challenging set of technologies. One is the UK's engagement with human fertilization, and the challenges of 'test-tube babies', IVF, stem cells and cloning. Over several decades, from the 1980s onwards, there was sober and open public debate about the issues, with an extraordinary level of both public and parliamentary interest. A regulator was established in 1990 – the Human Fertilization and Embryology Authority – which made a series of rulings that made often-cutting edge research possible, while also retaining public confidence.

But what is most striking about the HFEA is that it remains the exception rather than the rule.[12] It is rare in having translated the work of an advisory committee into a body with real power; rare in that it helped shape, and then be empowered by, a political and parliamentary consensus that marginalized its many enemies; and rare in that it retained public confidence. There was nothing comparable for the Internet, even as evidence mounted about its harms; nothing comparable for genetic modification or for the many other fields where technology was advancing at an extraordinary pace; nothing comparable for artificial intelligence, despite decades of hand-wringing.

Models such as the HFEA are useful prompts, even if they cannot be precisely replicated in other fields such as the metaverse or synthetic biology.[13] Institutions have to fit their context, and countries with polarized politics or strong religious institutions have to handle science in very different ways to ones that are consensual and predominantly secular.

We live in a very uneven world in other respects too, which limits the scope for standardized solutions. The US contributes roughly six times as much to greenhouse gases as the whole of Africa, whose population is four times greater. Countries differ by a factor of 42 in their neo-natal mortality rate; a factor of ten in the share of population with access to electricity; a factor of 50 in the share of population with access to the Internet; a factor of 2,500 in the number of scientific and technical journal articles per 1,000 population; and a factor of more than 1,000 in per capita energy related CO_2 emissions.[14]

The world is also uneven in its experiences of science, and far away from the standard story of a linear progression. Instead, eras are jumbled; old technologies reappear alongside new ones. Wood, once seen as primitive, is now a material of choice, for example for twenty-storey buildings in Scandinavia.[15] Most contemporary wars are fairly low-tech: the US was driven out of Afghanistan by Kalashnikovs and makeshift bombs, not missiles. Bicycles are the transport technology of choice in the world's richest cities, while the poorest ones improvise with homes made of corrugated iron, DIY water and electricity and mobile phones. Our aircraft typically were designed sixty years ago; our cars predominantly use internal combustion engines whose main designs date back 150 years. It's not surprising that perspectives vary and that generalizations mislead.

But, even in countries that are much more takers than shapers of science, there is no avoiding the need for politics to make decisions. The Internet is a good example. In its first decades it was largely seen as an opportunity (for prosperity or new ways of running public services): for most countries the only question was how to achieve more access to it and more services on it. Then belatedly it came to be seen as a threat (to childhood, morality and more) and in much of the world only thirty years after it became of part of everyday life were the first attempts made at more detailed regulation, for example of its effects on children's lives or privacy through requiring 'age-appropriate design'.[16]

Many believed that this offshoot of science could exist free from politics. Even though it originated in the military of the world's supreme superpower, and even though its central governing body, ICANN, was literally owned by the US Department of Commerce, it was hoped that this new infrastructure could be a vessel for pure freedom, entirely separate from states, politics or government. In the words of John Perry Barlow in the famous Declaration of Independence of Cyberspace in 1996, 'on behalf of the future I ask you of the past to leave us alone . . . you have no moral right to rule us'.

His motives were benign. But as Lawrence Lessig argued, 'liberty in cyberspace will not come from the absence of the state. Liberty there, as anywhere, will come from a state of a certain kind. We build a world where freedom can flourish not by removing from society any self-conscious control, but by setting it in a place where a particular kind of self-conscious control survives. We build liberty as our founders did, by setting society upon a certain constitution.'[17]

Many countries designed rules for the Internet that were a very long way from freedom, choosing instead to block, coerce and ban. But Lessig's fundamental point was right, and the fact that it took so long for politics to wake up says much about the gap between fast-moving technology and often sluggish governance.

The pattern was repeated a generation later with artificial intelligence: in the 2010s a flood of national strategies was published, promising to promote AI, and use it for economic ends, alongside another flood of haphazard attempts at self-regulation by the scientists involved. Only very belatedly, at the end of the 2010s, when AI was already built into many of the devices used daily by billions of citizens, did attention turn to the need for new rules and institutions to govern it with the kind of 'self-conscious control' that Lawrence Lessig advocated. Once again politics came in late, ambivalent about the facts and uncertain about what if anything it should do.

1.1 How science challenges political ideals

Science makes people powerful. It amplifies their senses, their mobility and their ability to think. Karl Marx was eloquent on its transformative power, writing in 1856 that 'steam, electricity and the spinning machine have been revolutionaries much more dangerous than citizens Barbes, Raspail and Blanqui' (who were leading radicals of his time). Yet, as Marx implied, science and technology are revolutionary not just in material terms but also in relation to ideas, as they threaten many of the most cherished assumptions of politics, the ideals that societies treat as foundational myths.

But what should the relationship of science and politics look like? In a famous letter Peter Kapitsa, a leading Soviet physicist, wrote to Stalin, asking for just such respect. 'There was a time when alongside the emperor stood the patriarch [but] the church is becoming obsolete . . . but the country cannot manage without leaders in the sphere of ideas . . . only science and scientists can move our technology, economy and state order forward . . . sooner or later we will have to raise scientists to the rank of patriarch' and without that 'patriarchal position for scientists the country cannot grow culturally on its own, just as Bacon noted in his *New Atlantis*'.[18]

Over the following decades scientists did become almost like patriarchs in many countries, with figures like Albert Einstein and Stephen Hawking seen as ultimate sources of wisdom. In the USSR, however, Kapitsa's suggestion was not taken up. Instead, he and his contemporaries were cowed into submission. Some became dissidents, like Andrei Sakharov and many others who had worked on nuclear weapons and realized the appalling power they had in their hands, and were then made victims of a corrupted science of psychiatry used to show that dissent and madness were inextricably linked.

Most states were content just to use science. But a few tried to subordinate it to ideology. One of the most famous examples is the promotion of Lysenkoism in the USSR in the 1930s, which was presented as a version of biology more in keeping with Marxism than mainstream Mendelian genetics, which was denounced as a bourgeois pseudo-science. (Lysenko remained as head of the USSR's Institute of Genetics until 1965, when attacks from Kapitsa and others finally forced him out.) This assertion of political science not only stunted science in the USSR but was also exported. Mendelian geneticists in newly Communist China were banned too.

Politicians couldn't help notice the affinities between scientific ideas and political ones. Atomism and individualism have a rough affinity, a view of the world as made up of separate particles. Against them stand ideas of holism, a view of the world as made up of wholes. This can lead in many political directions: the Nazis believed in what they thought was a biological view of society, seen as an organism, and many ecological and socialist strands of politics have also emphasized collectives, communities and systems. Darwinism's emphasis on the survival of the fittest could justify capitalism, while the global science of climate could support a comparably global view of good governance. James Lovelock and Lynn Margulis' concept of Gaia, seeing the Earth as an organism with self-stabilizing capacities,[19] can chime with a post-humanist politics. These alignments are loose at best, but our brains can't help but make links across different domains.

The subordination of science to ideology might appear to have been a mid-twentieth-century quirk. But it has also been seen in India with

Narendra Modi's recent promotion of a distinctively Hindutva science, strongly focused on cows, part of a commitment to 'Atma Nirbhar Barat', or self-reliance. History is rewritten so that, for example, Pythagoras' theorem was first invented in India (even though, ironically, historians now believe that Pythagoras' theorem was well known in ancient Mesopotamia, and China, long before India or Greece). Even genetics is presented as an Indian discovery. The character Karna from the Mahabharata is reinterpreted as an in vitro baby, which, Modi claimed in a widely reported speech in 2014, 'means that genetic science was present at that time'.

In the writings of Rajiv Malhotra, an influential commentator, science has to be fitted into a Vedic context. Modern science is *smriti* – a human construct based on sensory knowledge and reasoning – but underpinned by the Vedic *shruti*, the 'eternal, absolute truth unfiltered by the human mind or context'.[20]

In one plausible view, such reinterpretations of science may become widespread. In 2016, for example, the Chinese government began a programme to promote Chinese medicine, and two years later persuaded the WHO to include Chinese diagnostic categories in its International Statistical Classification of Diseases. Distinctively Arab, African, Turkish or Islamic science may be encouraged in the context of nationalist ideologies. Such combinations could enrich science – as in the case of Tu Youyou, from the Chinese Academy of Traditional Medicine, who won the 2015 Nobel Prize for her treatment of malaria. But there is an equal risk of spreading delusion and myth-making, obscuring facts and clarity in clouds of wishful thinking.

So, rather than seeking too much affinity between civilizational traditions or political ideologies on the one hand, and science on the other, a more helpful perspective sees science as naturally in tension with political ideas. Modern liberal democratic societies take for granted that science challenges the myths of other types of regime. It shows how nationalisms often rest on spurious histories or claims of racial purity that are upended by genetic knowledge. Democracies like to feel superior to the arbitrary and implausible beliefs of polities based on organized religion or of monarchies founded on false claims to their own historical provenance.

1.2 Science and liberal democracy

But the founding myths of the liberal democracies themselves, and their most cherished beliefs in freedom, equality and democracy are not supported by, or aligned with, scientific knowledge in any straightforward way.[21]

Take freedom. We sometimes think of freedom as the absence of restraints (for example on speech) or the absence of want, or fear, or in a more positive version, as the possession of capabilities. Science can certainly amplify these kinds of freedom.

But science also limits it. Indeed, if we ask what the opposite of science is it may not always be superstition, myth, unreason and ignorance, or the deliberate spreading of falsehoods.[22] Sometimes the opposite of science is freedom, because accepting science means accepting that we cannot believe whatever we want, that we have to accept the accumulated, shared view of what is true and what is false. Accepting science means that we cannot act without limits, because science tells us so much more about the possible effects of our actions, whether on our own bodies, the happiness of our friends and neighbours, or the prospects of future generations.

A world infused with science is a constrained world. It chafes. It tells us unwelcome facts, surfaces unwelcome problems, from climate to diet, and stands against free will, whether impulse, gut or intuition and our right to believe what we want: all are viewed with disdain through the eyes of science. To take just one example: replacing 80 per cent of beef with mycoprotein would eliminate about 90 per cent of forest loss.[23] But how many leaders want to confront their populations with such a stark message from science?

In a world infused with scientific knowledge freedom is not just an absence of restraints but also something that has to be engineered. This has always happened through constitutions and laws. But it is increasingly a matter of science too. How far should we have rights to be forgotten, to be ignored when we use the Internet or travel around a city leaving traces of every step we take? What is our freedom not to have our air polluted or new substances in our food? Conversely how free should we be to decide what substances we put in our bodies?

Or what of equality? Aristotle believed that inequality damaged politics: the community needs to share things in common to hold together.

Scientific knowledge can be reassuring for anyone who believes in the fundamental equality of human beings. It shows the remarkable genetic similarity of different races, or how many inequalities rest on arbitrary patterns of inheritance and luck rather than merit and virtue. But scientific observation also disturbs. It shows the very uneven distributions of physical strength, health, intelligence, aptitude, creativity, empathy, beauty and more. These do not necessarily undermine the desirability of embedding equal respect and equal rights into laws and constitutions. But these are best understood as chosen counters to inequalities rather than as emanating from any more fundamental equality that is rooted in science.[24]

Many advocates of equality favoured eugenics and birth control, believing them to be scientific tools for a fairer society, but justified precisely because people are so unequal. Social democratic Sweden, for example, sterilized some 60,000 Swedish women between the 1930s and 1970s. Keynes argued for legalizing contraception because 'to put difficulties in the way of the use of (contraception) checks increases in the proportion of the population born from those who from drunkenness or ignorance or extreme lack of prudence are not only incapable of virtue but incapable also of that degree of prudence which is involved in the use of checks'. A generation later Indira Gandhi introduced forced sterilization into India for similar reasons.

Or take democracy. Mass democracy rests on the idea that the people are best placed to understand their interests and to decide who should rule them. Yet science is by its nature hierarchical: not all opinions are equal. That a view is popular does not make it right. Moreover, science has shown just how distorted our decision-making heuristics are: how capricious, perverse, self-defeating and easily manipulated we all are. Much recent political science has taken pleasure in demolishing the happy myths. Christopher Achen and Larry Bartels argue that a 'folk theory of democracy' in which the people are always right and the job of politicians is only to follow is profoundly misleading. Voters choose which parties to vote for more because of identities or loyalty than policy positions and then adjust their views to fit their loyalties rather than because of any facts.[25] They are easily swayed by demagogues and misinformation. They misunderstand basic facts. Moreover, the system hardly reflects their beliefs. A longitudinal study in the United States concluded that 'the preferences of the average American appear to have only a minuscule, near-zero, statistically non-significant impact upon public policy'[26] (The rich, it turned out, have far more influence.) Behavioural science has shown again and again that the heuristics used to guide decisions are far from rational. Of course, this same science should make us even more sceptical of autocracy or monarchy, which amplify human vices to an even greater extent than democracy. But holding strongly onto the founding myths of democracy requires us to suspend and ignore much of what science tells us.

Or take the idea of a common good. Our politics rests on the notion that this is something that can, in principle, be defined and discovered. But in innumerable cases that common good is hard to define or pin down. An example is the contentious issue of Genetically Modified Organisms in agriculture, which I worked on when in government and later experienced at close hand as a laboratory near where I lived became a target for campaigners. GMOs may have benefits for food production, offering higher yields and less need for fertilizer. However, they can bring risks if they spread. They may have benefits for health (for example, if rice is enriched with vitamins) but also dangers if they make crops vulnerable to new diseases, or threaten people with novel toxins. The Food and Agriculture Organization (FAO), like many governments, claims that the benefits outweigh the risks.[27] Yet the more we look, the less obvious it is what the public or common good is, or, to be more precise, the more we discover that public goods are multiple not singular. What matters to people is multi-dimensional, complex and shifting: it includes a safe environment, cheaper food, healthier food and much more, and only intensive dialogue can bring out how these relate to each other and to the actual potentials of new technologies. Indeed, the common good turns out to be less a thing, a noun, and more a verb, a process of discovery.

1.3 The drive for sovereignty and its limits

Richard Sennett recalls an encounter with his old teacher Hannah Arendt on the streets of New York in 1962, right in the middle of the Cuban Missile crisis. Arendt informed him that the horrors of the bomb proved that you couldn't trust engineers and scientists when left to their own devices: politics had to intervene with, indeed supervene over, those who crafted our technologies. But how? Arendt was commenting from the perspective of sovereignty: the idea that we find the fullest expression of freedom and citizenship as active participants in the exercise of sovereign power, and through politics, which sits above all else.

As I will show, these ideas of sovereignty were adapted from kingship – the claims of monarchs to represent God in the governance of the world, without constraint or limits. Adapting these myths so that sovereignty was vested in the people was useful during the transition from monarchy to democracy, and these ideas continue to serve us wherever power is hoarded or claimed by rulers without much virtue.

But the myths no longer serve us so well. There has always been an uncomfortable tension between the constitutional idea of sovereignty – in which the people rule – and the actual life of governments, which rule over the people. But the myths sit even more uneasily in a world saturated with knowledge of all kinds, and a world in which science is pervasive and deep, because the obvious question arises, repeatedly: why should the views of the people, or an autocrat, be legitimate if they are at odds with what we know from science, if collective power is at odds with collective knowledge? Moreover, why should the views of the people today override what collective knowledge tells us about the damage we may be doing to the natural world on which future generations depend?

These dilemmas become even sharper when we acknowledge just how much science has amplified human sovereignty, making us collectively godlike. As the Nobel Laureate Paul J. Crutzen and his co-author C. Schwägerl put it, as they argued that we are now in the Anthropocene, 'it's no longer us against "nature". Instead, it's we who decide what nature is and what it will be'.[28] Science confers power but also reveals the risks of abusing that power.

It follows that any assertion of political authority over science has to involve a partial suspension of that sovereignty, a self-limiting move that is the only protection against illusion and delusion, as well as exploitation. For individuals, wisdom and maturity involve recognizing the limits of our knowledge and power, our dependence on others. The same is true of democracy. A mature sovereign people realizes the limits of its sovereignty, its dependence on others including the lumbering machineries of everyday government, and in particular its dependence on collective knowledge. A subject that is self-aware knows its limits as a subject. This is as true of a sovereign individual as it is of a nation. If it believes in fantasies of autonomy or self-sufficiency it will pay a price, precisely because these are, to a large extent, fantasies.

A paradox of democracy is that the more self-aware the sovereign people become, the more aware they become of the limits of their own intelligence, its bias and distortions, problems and gaps, and its reliance on other kinds of intelligence. Something very similar happens in our own individual lives when we delegate to others who know more than us, to build our homes, fix our cars, or cure our diseases. Sovereignty, in short, does not mean self-sufficiency or autonomy but rather a mix of freedom and humble dependence, and wise sovereignty limits itself.

2

What is science and how does it connect to power?

So what is the science that could be influencing political decisions, or being guided by politics? There are many answers. Science is about flashes of individual genius but also about huge organizations. It is about serendipity but also about bureaucracy and process. It is about the public good but also about profit.

Some focus on what it knows: Herbert Simon called science humanity's library of facts about the world and explanations for those facts.[1] Some focus on the ethos of science as its essence. That ethos includes the spirit of discovery and organized curiosity, and much more, with questions fuelling answers and answers prompting new questions. Heidegger quotes Silesius: 'the rose is without why; it blooms because it blooms', whereas the 'principle of reason' is constantly to ask why, and we might add 'to develop powerful methods of verification'.[2]

Others focus on method as the defining feature of science: the use of experiments, data and critical review by a community of peers. In the words of Richard Feynman, 'if it disagrees with experiment, it's wrong. And that simple statement is the key to science. It doesn't make a difference how beautiful your guess is, it doesn't matter how smart you are, who made the guess, or what his name is. If it disagrees with experiment, it's wrong. That's all there is to it.' These make it possible to distinguish pseudo-science (which can't be falsified) from true science that can.[3]

Others like to point out that although science has the appearance of being a definable domain, with clear principles and boundaries, in reality it intersects with many others –the worlds of lawyers; accountants; investors and businesses; research managers and funders; publishers; the buildings, machines and devices that help to sense, measure and distil. All are parties to the world of science and without them it couldn't function.

To understand the relationships of science to power it's useful to separate out three distinct types of activity that are brought together under the umbrella term 'science', each of which has distinct political and moral characters, and different implications for power.

2.1 Observation: trying to see the world as it is

The first of these is sensing and observation – the accumulation of means of observing the world as it is, not as we would wish it to be. These range from measuring sticks to the James Webb Space Telescope, sensors of pollution, air and sounds, to probes, colliders, thermometers and microscopes – and often in history observation preceded breakthroughs and novel theories. Seeing microbes in wounds prompted work on how to sterilize them. Seeing the interior of a cell (through X-rays) prompted the idea of a helix in DNA. Louis Pasteur's interest in how fermentation could lead to better wines led him via observations to new insights on chemical transformations that led ultimately to pasteurization.

The many tools with which science can now observe have extended our senses and they are in part a collective property. They allow us to monitor deforestation rates in the Amazon; particulates in the air of a city; patterns of behaviour on city streets. Our ability to zoom in and out, from the microscopic to the cosmic, gives us dramatic insights into the way the world works.[4]

Such observations would be enough to transform much of how we live, and they alone change balances of power. Superior observation gives you leverage over others, which is why states have invested so heavily in means of observation and intelligence, from intercepting letters to trawling digital communications, lookouts to satellites. But even observations can be contradictory and ambiguous: what we choose to look at determines what we see. So if, for example, in a pandemic there is intensive observation of infections and economic activity, but little of levels of anxiety and loneliness, this is bound to shape the decisions and actions that then follow.

2.2 Interpretation and sense-making

Next comes sense-making: interpreting the signals with theories of all kinds that have explanatory power, that explain why the rose grows, or why we see it as red, or why it suffers parasites. This is the science of theories – the theories that explained heat and power; the theories that explained cosmic patterns, with relativity and quantum physics; the theories that explain how two metres of DNA in every cell in our body shapes what we are. These are the results of creative imagination and what Charles Peirce called 'abduction', intelligent guesswork: the suggestion of hypotheses and explanations that can then be tested. They are best when parsimonious: simple enough to explain but not simplistic. Nabokov commented on the psychological satisfaction of such robust interpretations: 'I cannot separate the aesthetic pleasure of seeing a butterfly and the scientific pleasure of knowing what it is.'

That pleasure derives from the combination of observation, experiment and interpretation. Experiment must have been part of human life from the earliest times: experiment to discover what to eat, what to burn, how to hunt, how to survive. But it is surprisingly recent in the history of science and research, which relied more on deduction and reflection until the seventeenth century.

A world in which we look out through more reliable interpretations and theories is very different from one without. A society with robust interpretations is less surprised by heart attacks, seismic eruptions or droughts, and less surprised when an army with rifles and missiles defeats an army that relies on gods or magic to give it victory. Here, too, successful interpretation changes power dynamics. Nations with more accurate interpretations and theories defeat ones with less accurate interpretations of the world.

But interpretation too can be conflicted and ambiguous. People can look at the same data and reach different conclusions, despite an array of sophisticated methods for inference and reasoning.[5] Interpretations can embed biases of all kinds and they can both amplify power and challenge it.

2.3 Action

After interpretation come making and doing: action on the world to produce new materials or products, where science becomes technology and engineering, and changes the world by becoming the environment in which we live.[6] Science is realized through technology, just as written music is realized through a performance. Here we come to the development part of research and development, which demands a mindset of sensitivity to materials, contexts and costs that sometimes looks very different to the expansive mindset of scientific discovery.

This is where science affects our daily life. The diets recommended for meals; the mobile phone using GPS; the pills we take to manage a chronic condition; the electric bus we take to work that depends on sophisticated storage and distribution; the synthetic materials in our shoes; perhaps the sleeping pill we take last thing at night or the lights designed to slow our brains down. Almost every sphere of life is in some way touched by organized science and ahead of us lie decades when new materials, such as incredibly strong and flexible spider silk protein or adapted myceliums, may transform our environments as the stuff around us becomes smarter, able to regulate our temperature or protect us from accidents.

These are examples of applied science, the collective gift on which our lives depend when we are sick, when we need to travel, when we want to solve problems, gaze at the skies or try to make sense of the natural world around us. Without it most of us are helpless. Yet the effects of technology are rarely simple: there is no linear relationship between new technology and benign outcomes. More often, some people are empowered, others disempowered, some enriched, others impoverished.

The traditional view of science painted a picture of a logical progression from abstract theory through to practice. But an alternative view sees much more interplay between practice and theory. In the words of the classic history of twentieth-century science, 'no meaningful science has been generated that cannot be identified with a working world origin':[7] an engagement with live, real problems sparked the new insights. Others, too, emphasize the dance between new questions and new answers, in which the distinction between research and development, science and technology often breaks down.[8]

Technology is, in any case, a slippery word, used for things (like smartphones), processes (like smelting steel) as well as for the knowledge that lies behind them. In each of these meanings it is seen to evolve through its own logics, dependent on the paths already taken and what already exists.[9] Technologies create building blocks that can then be assembled and combined in novel ways,[10] with each step opening up new adjacent possibilities, and novel ideas spreading through a process of attraction.[11] These logics then carry with them implications: as Melvin Kranzberg commented, with a hint of irony, technology is not good or bad – or neutral.[12]

Here the links to power are very obvious. Technological superiority powered many empires, from China and Persia to Rome and Britain, allowing them to feed large cities or communicate over vast distances, just as technological capability now allows many regimes to remain in power despite mass opposition. If we use Michael Mann's framework for understanding the different forms of social power, all were shaped by technologies: tools for coercion and violence; for finance; for administration; and for persuasion.[13]

These three dimensions of science, observing, interpreting and acting are sometimes independent of on another. We can act on the world without very good observations and theories (though we are probably more likely to make mistakes). Personal relationships and career choices are often examples. And we can choose simply to observe but not to act (as most astronomy has always had to). Technology can also evolve without science and did so for much of human history, with inventions from gunpowder to optical lenses happening without much scientific theory to support them.[14]

Science moves in haphazard paths, rarely in straight lines, partly because each step from observation to interpretation to action involves a leap. The familiar story of the discovery of penicillin sometimes glosses over the fact that Alexander Fleming gave up before others went onto manufacture it at scale. CRISPR gene-editing technology came out of work done by a Spanish researcher investigating organisms in stagnant water, who discovered repeating sequences of DNA, and then benefited from contributions from researchers at a Danish yoghurt company interested in understanding the bacteria in starter cultures. Again and again, we find stories of serendipity, lateral moves, conflicts, bad blood and surprising combinations in the true histories of science, which bear little relation to the fairy-tale stories of individual genius and foresight that we learn as children.

But it is the combination of observation, interpretation and action that we usually mean when we talk of science, and it is the three together that make up most of the impact of science in the world. This is why I argue that these are also the three tools that politics needs to apply to science itself: better observing what it does, interpreting its patterns, benefits and risks and then acting in the light of that interpretation to accelerate, slow down or block different scientific and technological pathways.

2.4 *The collective nature of science*

That task is made easier by the open and collective nature of contemporary science. The word 'gibberish' is thought to come from the eighth-century Arabic alchemist Jabar ibn Hayyan, who was so fearful of being executed for black magic that he wrote his findings in the most obscure way possible, with the result that almost no one could understand him. For most of history science was secretive, arcane, obscure and often considered indistinguishable from magic.

Modern science by contrast combines observation, interpretation and action in forms that collectivize the knowledge gained – the observations, the theories and explanations, the means of action in the world – and institutionalizes them in labs, centres, disciplines, funds and stored memories.[15] As a collective, science polices itself, as happened in 2018 when a Chinese scientist, He Jiankui, announced the birth of twin girls with edited genomes, and was met with a storm of disapproval and sanctions (and a three-year prison sentence).

This open and collective nature was understood early in the history of modern science. Joseph Glanvill was one of its first theorists, arguing in the 1660s that 'free and ingenious exchange of the reasons of our particular sentiments' is the best method of discovering truth and improving knowledge, and suggesting that science like nature works 'by an Invisible Hand in all things' (a century later Adam Smith borrowed this metaphor to explain how economies work).[16] Yet this, the increasingly collective nature of science, is often missed in stories of individual genius, whether Newton sitting under an apple tree or Einstein writing at night after his job at the patent office. But the more we know, the more collective science looks, dependent on networks of collaborators, supporters and colleagues.

A good example of the collective nature of discovery is DNA. The simple histories still say that DNA was discovered by Francis Crick and James Watson. Yet by the early 1950s it was already known that long polymers in the nucleus of cells carried genes; it was known that they were made up of four nucleobases, with always equal amounts of cytosine and guanine, and adenine and thymine. What wasn't known was the structure, which Crick and Watson inferred with the help of a photograph taken by Rosalind Franklin (helped by Maurice Wilkins who shared the Nobel Prize but was largely forgotten). Darwinian evolution was similarly a collective discovery, and to his credit Darwin acknowledged the parallel work by Wallace and earlier work by William Charles Wells and Patrick Matthew decades earlier. That so many inventions have appeared simultaneously, but independently, in multiple places, confirms that science is better understood as a living, shared community[17] and this collective, networked characteristic of science is even more pronounced now. Between 2014 and 2018 there were 1,315 scientific papers with more than 1,000 authors, double the rate for the previous five years.[18]

These numbers could explode as citizen science in all its forms continues to grow.[19] Millions volunteer to help platforms like Galaxy Zoo (finding and classifying galaxies) or Patients like Me (sharing insights on health conditions) and it seems that the more people take part in science the more they understand it and have confidence in it: in one possible future science becomes ever more integrated into daily life, as millions take part in observing, experimenting and interpreting everything, from symptoms in health to the state of flora and fauna. Geography is also part of this story. If scientists migrate to places where a lot of science is happening, they become much more productive. The context and milieu matter as much as the genius.[20] The collective allows the individual to flower.

If science itself has become more collective, so is its political quality ever more a function of the connections made between the different strands of science mentioned above: observation, interpretation and action. As an example: it is scientific observation that allows us to see the dramatically different ecological footprints of different social groups, some brazenly predatory in their use of huge quantities of materials and their vast carbon outputs; others modest and walking lightly on the Earth. These observations then in turn prompt us to ask why and how such inequalities arise, and what role the material realities of science and technology play in amplifying them. Finally, they then raise the question of action – of what can be done to rein in the predators and reward those who give, and work hard. This kind of loop, repeated a thousand times, generates the politics of our times: a science that sparks, illuminates, questions and then loops back on itself in a pressure for action.

2.5 The idea of a scientific state

Shortly after his death, Francis Bacon's book *New Atlantis* was published.[21] It suggested a new way of looking at science, though several decades passed before its significance came to be recognized. The book describes a utopia based around research and is perhaps the first explicit vision of a polity with science at its core. A central role in *New Atlantis* is played by the House of Solomon, which organizes research '. . . to discern between divine miracles, works of nature, works of art, and impostures and illusions of all sorts; the knowledge of causes, and secret motions of things; and the enlarging of the bounds of human empire, to the effecting of all things possible.'

The *New Atlantis* pointed to a changing relationship with power, with science steadily filling a space once filled by magic, astrology and arcane knowledge. It also points to a future where science would find itself in competition with other kinds of power, and other kinds of capital, in a fight for prestige, recognition, resources and freedom, which required it to mobilize many weapons: its own usefulness and successes; argument; narrative; imagery; myth; emotion and more.

Bacon's book also showed that the interest would run both ways. Science often portrays itself as innocent, the disinterested pursuit of curiosity. But it has also always needed power to provide it with its tools, even just to observe. It might need a monarch to provide an observatory or libraries, or a large state to fund underground colliders or to mobilize the vast resources needed for a space mission. Such state patronage has a long history. In the US, for example, Jefferson funded a 'Corps of Discovery' to explore the West; a few decades later the land-grant universities underpinned dramatic innovations and improvements in agricultural productivity, with hybrid corns and new ways of combatting pests; and the funding for Samuel Morse's first telegraph came from Congress.

Collective knowledge required collective commitment, and the support of power – and it still does two centuries later. Science needs money, often vast amounts. The Manhattan Project cost some $26bn.[22] In the USSR spending on science was lavish, appropriately for a state that believed itself to be based on scientific principles. Big projects in physics are extraordinarily expensive, and examples of the profound uncertainty involved in any activity on the frontiers of knowledge. The Large Hadron Collider for example cost $4.75bn and new colliders may cost $1bn a year. It's entirely possible they will open radical new possibilities and create whole new industries, just as the earlier advances in electricity, electro-magnetism and quantum can plausibly be described as a line linking abstract high theory and everyday results, though no one can be certain, and some argue that 'while the cost of these colliders has ballooned, their relevance has declined'.[23] What isn't in doubt is that only states have the capacity to generate such huge resources.

2.6 The political character of science

There is a large and long-standing literature on the political character of science and the technologies it supports.[24] Political priorities shape both what gets developed and what gets used – whether the siege engines of medieval times, the Zyklon gas used in Auschwitz, the atom bombs dropped on Japan, or the Concorde aircraft, fuelled by French and British public money, designed to speed the travel of the very rich. Politics can be embedded into AI algorithms making decisions on probation that are trained on the decisions of white judges or into tools like the DSM manual (the dominant guide to psychological treatments in the US), which categorized homosexuality as a mental illness until the early 1970s. Technologies like nuclear power can tend to favour centralization and a security state, while others, like solar panels, can do the opposite.[25] Decisions over grand geo-engineering projects to reduce planetary warming – like dropping sulphates into the stratosphere to reduce sunlight hitting the Earth – cannot avoid having potentially vast political consequences and a dramatically uneven distribution of risks and rewards.

The shape of science also inevitably reflects power distributions. For example, in healthcare neo-natal conditions, respiratory infections and nutritional deficiencies represent a high burden in poorer countries. But research on these is underrepresented even in these countries, and much more so globally. By comparison some 40 per cent of global health-related research focuses on cancers, which accounts for only five per cent of the disease burden in low and lower-middle-income countries.

In all of these ways politics is enmeshed with science and vice versa, which may be why every modern political project has tried to mobilize science. For nineteenth-century liberalism, it was important to show that telegraphs and steam ships and knowledge of new materials could expand freedom. For the students trying to modernize China in the 4 May movement, 'Mr Science' and 'Mr Democracy' were the twins who could save the country from chaos and colonization. In the words of Chen Duxiu, writing in 1919, 'only these two gentlemen can save China from the political, moral, academic and intellectual darkness in which it finds itself'. Almost exactly a century later, in 2018, China surpassed the US in the scale of its scientific production (though it had, of course, dispensed with the services of the second gentleman).

Lenin advocated 'Soviet Power + Prussian railroad organization + American technology + the trusts = socialism' (and in the 1960s the Soviet Communist Party promoted the notion of a 'Scientific Technical Revolution' founded on automation in the same spirit). For late twentieth-century neoliberalism it was vital to claim the Internet as a platform for global exchange, while twenty-first-century Green politics harnesses the scientists' warnings of climate change and biodiversity collapse.

But science has no natural affinity with any particular philosophical or political stance, or rather it can align with any. Excellent scientists have been Nazi, communist, Catholic, Muslim, Quaker and Mormon without any great struggle or internal conflict. One historian, for example, commented that a majority of 'Germany physicists like biologists, welcomed Nazi rule'.[26] Rudolf Hess called Nazism 'applied biology' and his colleagues drew on the prevailing interest in germs, hygiene and purity, convinced that their ideas aligned with the leading-edge thinking of their times. Eugenic ideas were popular among socialists as well as conservatives: it seemed modern, rational and scientific to want to influence patterns of inheritance. Bernard Shaw, for example, wrote that 'the only fundamental and possible socialism is the socialization of the selective breeding of man' while Bertrand Russell suggested that governments should issue colour-coded 'procreation tickets', and that people should be fined if they had children with a partner with a different colour, to ensure that the genetic quality of the elite was preserved and not diluted.

There were times when it appeared that authoritarian regimes would be inherently better at harnessing science. Nazi, fascist and communist science would outperform the democracies because of their single-mindedness and their will to win. They would not be constrained by ethical worries and by channelling scientist's energies into a few national priorities they would avoid the dissipation of effort that follows from giving scientists too much freedom. Then, after the Second World War, a mirror view took hold that cultures of hierarchy, fear, constraint would so inhibit free exploration that the opposite would result. Without the intrinsic motivation of curiosity science would stall. Without freedom to deviate, science would be led down blind alleys.

But science has flourished in very varied environments so long as scientists were given resources and some space to work. The advance of rocket technology under the Nazis or supercomputing in contemporary China should put paid to any confidence that the logic of science is limited to particular ideological homes. Instead, it can sit as a self-referential system of logics alongside almost any other body of thought, from revelatory religion to the most reactionary politics.

Just as it can align, it can also ignore. Recent years have brought attention to the gaps, the missing elements, with the concept of epistemic injustice capturing the notion that shared epistemic resources are allocated unevenly, blocking out the interests and values of some groups in favour of others.[27] Science has certainly tended to be a tool for the dominant, ignoring or suppressing the perspectives, values and needs of the powerless and the 'muted', a topic now being studied in the discipline of 'ignorance studies', which addresses the unknown and unseen.[28]

It would be comforting to believe that the maximum openness and transparency is always for the best, and that there is a natural home for science in the most open democracies.[29] But science has advanced in many places without much space for freedom of speech and inquiry and in some democracies research has become more secret in response to heightened geopolitical competition, echoing a history in which a large part of science has always been jealously guarded for reasons of security. Indeed, secrecy, and lack of accountability, sometimes allows for more risk and more wild ideas. The US intelligence agencies, for example, employed many scientists and technologists who were given free rein in everything from attempts to develop city-destroying Tsunamis, to bionic cats, nuking the moon to the vast industrial scale surveillance methods of the NSA.[30]

The conclusion we should draw is not that there is a strong affinity between any kind of science and any kind of politics, but rather that this is a question of design: that any society can seek to design institutions, funds and rewards to align science to what it cares about. Indeed, it has no choice but to design if it wants any science to happen: the question is not whether it designs, but whether it designs well.

PART II

How States Have Used Science

3

The ages of techne and episteme

To understand the contemporary dilemmas of science and power we need to understand the paths that brought us here, and how the power of states and the power of science co-evolved, slowly over thousands of years and then very rapidly over the last two centuries. In what follows I describe distinct phases. Each grew out of the previous one but added a new layer on the previous layers, rather than superseding them. These can be understood through Aristotle's distinction between three different ways of thinking. Techne is the practical knowledge of things. Episteme is the logical thinking that applies rules, and phronesis is practical wisdom, the sister of sophia.

For a long period, the state's engagement with science was essentially about techne: building cities, irrigation canals and channels, war machines, roads and walls. These actions were often dressed up in ritual of all kinds. But they were functional, utilitarian and practical, and vital to any state seeking to exercise power at any scale.

Then in a much more recent phase, which begins in the nineteenth century, science became more organized, and more defined by episteme. It came to be seen by states as a new way to amplify their power, both against other states and against their citizens, and, sometimes, to help their citizens to thrive: this is the era of sophisticated technologies of warfare, from ships and rockets to tanks and planes, and of deliberate state policies to build up institutes, labs and centres that could harness science in the national interest. The state could specify desired ends, and hope that science would come up with the means.

Then, in a most recent phase, which we can roughly date to 1945, just as the tools of the second phase reached their apogee, with far greater spending on science, the issues became more complex. Science turned out to generate risks as well as possibilities: weapons that could wipe out humanity; computers that could undermine societies; pollution that could kill. Science, the servant of power, turned out to be more ambiguous than had it appeared previously, its effects less predictable, and so states had to find new ways to regulate, assess, ban or block. Ends and means became interwoven, and in this phase, phronesis became ever more essential: the wisdom to see and think in multiple dimensions. This is the phase we are still in, struggling to cope.

3.1 Engineering in the service of power

For a very long time the relationship between science and states was essentially about things: engineering in the service of power. A good example is the Dujiangyan, a very old system of irrigation that's still used today in Sichuan in western China. It was built in the middle of the third century BCE, under the leadership of an engineer, Li Bing. Before then, the river flooded repeatedly, often with disastrous effects, as melting ice from the mountains came down into the slower-moving river. A dam could have been an option. But it was important to keep the river open for trade. Many tried to dream up solutions, but without success.

In the end the solution, an extraordinarily bold one, was a levee that could redirect the flow through a mountain to a drier plain. Achieving it required an army of thousands of workers to make fires to heat the rocks and then cool them down repeatedly so as to crack them open. Once completed, the river never flooded again. Sichuan became the most productive farming area of China. Indeed, life was so prosperous that it was claimed the region's relaxed attitude to life resulted from this feat of engineering.

The triumph was explained to the people in a very unscientific way, through an elaborate ritual that symbolized defeating the river god. But here we see one example of many of a partnership between science and the state. The levee in Sichuan was an instrumental mobilization of knowledge to solve a problem that then enhanced the prestige and wealth of the state.

Humans have always relied on knowledge of plants, animals, seasons and weapons, and leaders have always relied on others with more specialized knowledge than themselves. But the early states went a step further, trying to be systematic orchestrators of specialized knowledge as well as ritual and much else. What mattered most for them was knowledge of food and water, including the management of irrigation from

Mesopotamia and Ancient Egypt to China. That knowledge didn't have to be centralized. In Egypt, networks of earth banks and regulated sluices managed the annual surges of water bringing silt from the Ethiopian highlands. Measuring devices were installed along the river to help predict each year's levels, since too low a level led to famine and too high a level to disastrous flooding. Yet the management was mainly local, not dependent on the central state.

The cities of Sumeria were more centralized and invented arithmetic for their own administrative needs. They monopolized grain production and required their people to work for it, and so needed ways to represent time worked and the allocations of food. Innovative representations and abstractions made it easier to govern (and the first writings are lists of things, not of gods).

Sophisticated urban design, water systems and sewage can be seen in ancient cities from Mohenjo-Daro to Tell Brak, Uruk and later Babylon. The people in charge of these projects observed, generated models and experimented, in parallel with the more abstract knowledge of astronomers and astrologists. Their builders were supported by states able to generate and deploy surpluses.

This legacy of engineering in the service of power is with us today in the grand projects of twentieth and twenty-first century modernity, such as the Hoover Dam, or in the work of figures like Robert Moses, the great planner of New York, notorious for his social engineering as well as his physical engineering, building bridges so low that buses full of the urban poor could not come through. Engineers lead projects like the planned 170 km-long linear city of Neom in Saudi Arabia or Nusantara in Indonesia, or the Oresund bridge linking Copenhagen and Malmo that created a new economic region, or the remarkable railway that connects Chongqing to Lhasa, and runs at heights over 5,000 metres.

These physical projects are usually more than just physical: they also bear messages about power. But, perhaps precisely for this reason, these kinds of engineering innovation have often been the only spaces in which innovation could be permitted. Other kinds of novelty were suspect. During the early modern period, from the fifteenth to the nineteenth centuries, innovation was generally frowned upon, and even formally prohibited by laws (as in sixteenth-century England) if it went beyond engineering.[1] A spirit of speculative inquiry and discovery was as dangerous then as it may be in the many absolutist monarchies and dictatorships of the twenty-first century, which are as keen to demonstrate their prowess on the frontiers of material innovation as they are keen to stamp on any innovation in politics, society or thought.

3.2 Science as amplifier of state power

The more modern sense of science, and of its changed relationship to the state, is post-Copernican, and associated with the great acceleration that began in the late eighteenth century. Galileo had helped to usher in the science of observation and theory that uses maths to model the world and that relies on experiments as the best way to find and confirm new knowledge.

Newton complemented the emphasis of a previous generation on experiment and individual facts with a method that sought to generalize, with explanations more likely to come from calculation than observation, a period that brought 'the establishment of science as a distinct activity, to be controlled only by its own norms'.[2] But this, the reflective science of Newton or LaPlace embedded in a semi-autonomous republic of letters, was not immediately useful to emperors and kings, or to business. There were intimations of its future relevance, and later, in France, Condorcet advocated science for social emancipation and St Simon a utopia based on science, while in Germany, Humbolt advocated a science for social good, as 'the knowledge economy . . . became an important mechanism supporting the new political order', offering new ways to fight, to grow food or to dominate a population.[3] Mapping and measuring were also crucial: censuses, geographic surveys, the design of new metrics (like the metre and kilogram), surveys of the health of the population: all were ways to bring order to the world.

But the impacts were modest. Michael Polanyi claims that until the mid nineteenth century 'natural science had made no major contribution to technology. The Industrial Revolution had been achieved without scientific aid.'[4] Its successes were ones of engineering, innovation and experiment, techne rather than episteme.

It was not until the mid nineteenth century that more recognizably modern forms of science were becoming integrated into society and into the state. The latter half of the century brought the creation of formal institutions in England, France and Prussia that fused bureaucracy, science, engineering and politics, and was primarily directed at armaments, new battleships, guns, cannon as well as materials. In parallel there were the beginnings (notably through the ITU, the International Telecommunications Union, originally founded in the 1860s) of what later became a lattice of new global partnerships and institutions organized around technology.

The forms of this embryonic science state differed. In France, it focused around the hospital, museums and academies. As Bruno Latour explained in his study of late-nineteenth-century France, science became embedded in the daily workings of the state (what Latour called the 'pasteurization of France') with the creation of a huge socio-technical network focused on public health.

In Germany there were the research universities, industries founded on close ties to research, influenced by Friedrich List and others, mobilizing knowledge to fuel the blood and iron of Bismarck's ambitions, and create an industrial base that could challenge Britain. Meanwhile, in Britain the Navy played a decisive role, commissioning new armaments – such as the Dreadnought battleship – to protect Britain's dominance of the seas, though the state saw little need to intervene much in the technology of industries, whether textiles or chemicals, steel or rail.

A clear hierarchy of science had emerged, with physics as fundamental, on top of which sat chemistry, biology and the other sciences. It was physics that propelled forward electricity and the industries that depended on it, and it was physics that confirmed the idea of a science abstracted from and detached from everyday world challenges, but which could nevertheless be mobilized for the everyday.

By the early twentieth century a scientific-industrial complex was taking shape in all the leading countries and the First World War propelled it forward, bringing the first systematic mobilization of civilian scientists for war, in the UK, Germany and the US. They developed not just weapons, but also techniques like intelligence testing, accelerating science's evolution from a system linking a handful of clever individuals to large orchestrated systems employing tens of thousands.[5]

By the time of the Second World War, the Manhattan Project, to develop atomic bombs, was organized at a scale equivalent to the US car industry. The state proved adept at switching to a planned economy, fusing the business methods of mass industry to the needs of war. Vannevar Bush exemplified the shift. Before the US entered the war Roosevelt had asked him to mobilize all the talent of American universities, science and technology 'so that when we enter this war we can have as few casualties as possible and we can end this war as quickly as possible'. Later, having overseen a dramatic expansion of government support, and having written the classic text of twentieth-century science policy, 'Science: the endless frontier' in 1945, Bush persuaded the US government to create the largest ever system of research, with huge spending, a fair amount of autonomy and a fair claim to have fuelled decades of economic success.

But persuasive words were not enough. Bush had proposed to President Harry Truman a target budget of $120 million a year (over $1.6 billion in today's money) for medical research, natural sciences, defence and education. But the National Science Foundation launched in the early 1950s with a far smaller budget of only $3.5 million. It was only when the Soviet Union's Sputnik launch showed the scale of competition that the sums dramatically increased. Fear unlocked the fiscal gates. The money started flowing and has never stopped (with even larger sums channelled through the NIH in health, the Department of Defense and NASA).

Within a generation, a sum roughly equivalent to 2.5 per cent of US GDP was devoted to the Apollo programmes, an extraordinary sum for a nation much of which was still backward and poor. As a scientific project Apollo looks as impressive half a century later as it did at the time. But whether it was wise is far less clear, since it coincided with the start of a long period of stagnating incomes for most Americans and stagnant life expectancy. Even the most famous spin-off of the mission – the non-stick frying pan – was a myth: Teflon had been invented in the 1930s and Tefal's first non-stick frying pans went on sale in the 1950s. No wonder many asked why, if a man could be put on the moon, it was so hard to solve problems closer to home.[6]

Vannevar Bush exemplified the hopes that science could, to some extent, be planned, just like roads or steel. In the 1930s J.D. Bernal, one of the most influential Marxist scientists of his time, had advocated planned science as just one aspect of a more planned economy (and suggested turning the world into a human zoo where its 'inhabitants are not aware that they are there merely for the purpose of observation and experiment').

Even then it was clear that planning was difficult and that governments struggled with what came to be called the 'Collinridge dilemma'. The more governments knew about the likely implications and uses of a technology, the less chance they had to shape the technology, as its paths were already determined. Experience has further tempered hopes that there could be even a rough correlation between inputs and desired outputs. Nixon's war on cancer promised that 'the same kind of concentrated effort that split the atom and took man to the moon should be turned toward conquering this dread disease' and he hoped that this would be achieved by the time of the American bicentennial in 1976. A review published in 1986 concluded that 'some 35 years of intense effort focused largely on improving treatment must be judged a qualified failure',[7] while an updated version by the same authors in the late 1990s was titled 'Cancer Undefeated'.[8]

Old habits die hard, however, and in 2016, Joseph Biden, then Vice President, launched a 'Cancer Moonshot' initiative, which once again promised to end cancer. Many in the field felt ambivalent. Half a century had brought extraordinary advances in understanding of the complexities of cancer, its genetic aspects, and the roles of environment, but these had also made the political simplicities of 'wars', 'moonshots' and 'missions' look faintly silly. But perhaps this was the price to be paid for generous state funding.

The US's confidence that science and technology could solve all problems was mirrored in its great adversary, the USSR. Russia had a remarkable history of scientific breakthroughs, from the first punched cards for information storage in the 1830s, to the periodic table, the first electric streetlighting in the 1870s to Zworykin's prototype TV in 1911. By the 1950s it was devoting huge resources to research, a commitment that seemed to pay off with the world's first industrial nuclear power station in 1954, the Sputnik satellite and the successful journey of Yuri Gagarin into space. Yet, this extraordinary effort coincided with the beginnings of the stagnation that ultimately destroyed the regime. Indeed, Russia remains a lesson in just how hard it is to turn scientific prowess into state power. Today over 70 per cent of Russian science is funded by the state, the highest ratio for government spending on research worldwide, with funding focused on national security and the development of strategic industries, and 40 per cent of the budget financing heavy industry research. But Russia's economy remains predominantly reliant on hydrocarbons and raw materials. *Harvard's Atlas of Economic Complexity* ranks Russia 52nd of 133 nations and one study concluded that 'at a time when wealth depends to an increasing degree on knowledge, Russia does not have an effective system for converting its scientific capacity into wealth'.[9]

Russia also exemplifies the challenges of organization, dominated by the Russian Academy of Sciences (RAS) and a network of over 1,000 research institutes that accounts for almost 25 per cent of all research output. The direct government funding that supports over 4,000 research organizations and 750,000 research personnel has often undermined intellectual creativity, and indeed was challenged in the 2010s by the high-tech mega-centres such as the Skolkovo Innovation Centre and Akademgorodok in Novosibirsk – prompting a shift to funding science via competitive, independent peer-review, often including foreign subject experts.[10] But Russia remains a largely extractive economy and its relative failure was made painfully apparent after its invasion of Ukraine, when its weapons proved technologically inferior to the West's.[11]

By the early decades of the twenty-first century every state felt the need for science (though it is an oddity of mainstream economics that it has repeatedly forgotten the central role that states have played in science and technology). Some nations spent ever larger sums both for security and prosperity: three or four per cent of GDP in the cases of Israel, South Korea and Finland (not coincidentally small countries with hostile neighbours).

China is now the world's second science superpower, drawing on millennia of leadership in many domains. For many Chinese leaders it was baffling that Chinese civilization, despite its global pre-eminence, had missed out on the scientific revolution and had paid such a high price for its relative backwardness. Mao was determined to overturn this legacy of failure. It's been suggested that the gift of knowledge, intellectual property and assets from the USSR to China in the 1950s was the biggest ever such transfer from one country to another in human history. Drawing on this knowledge Mao set up programmes for nuclear weapons (with China's first test in 1967 shocking the world) and an Institute of Genetics in 1959 (having ditched Lysenko-ism as part of the break with the USSR). By the early twenty-first century China was competing with the US on dozens of fronts, landing on the dark side of the moon and spending almost as much in absolute terms as the US, and investing through thousands of 'Government Guidance Funds' providing venture capital investment for technology estimated at over $1trn.

No nation can now afford to fall behind even if it aspires to autarchy. North Korea uses organized crime, ransom-ware and theft of crypto currencies to finance its missile programme. Iran, a regime founded on faith rather than science, invests heavily not just in defence and cybersecurity but also in new fields. A surprising example is biology, a field where science becomes godlike and where you might expect a theocracy to resist new knowledge. Yet in 2002 Ayatollah Khamenei, Iran's Supreme Leader, issued a 'stem cell *fatwa*' which stated that experimentation with human embryos was consistent with *Shia* tradition and went on to congratulate the scientists who had produced human embryonic stem cell lines.[12] Afghanistan, where hundreds of scientists left when the Taliban regained control in 2021 and effectively shut the Afghanistan Science Academy, is an almost unique example of a state that appears to have no interest in science.

Later in the book I look at the potential democratization of science, and how it can be steered to broader public and social interests. But here I focus on the four primary purposes that have shaped the deep involvement of states in science: war, commerce, glory and domination.

3.3 Science for war

Almost every state has used science for war, or more specifically to gain an advantage over competing nation states. Vladimir Putin commented that whichever country leads in artificial intelligence 'will become the ruler of the world'. In previous eras gunpowder, siege machines, warships, rifles, Gatling guns, tanks, aircraft and submarines, and more recently drones, swarms and cyberwarfare are just some of the tools, using science, that states have commissioned to give them an edge. All become part of arms races that date back to the dawn of humanity: the spear that prompts the shield that prompts the stronger spear. City walls that prompt siege machines that prompt thicker walls. Missiles that prompt missiles to shoot down missiles. Cyber-attacks that prompt cyber defences that prompt smarter cyber attacks, and cognitive attack that prompts cognitive defence.

As indicated earlier, the First World War was the first when science was mobilized on a large scale to create tanks, aircraft, chemical weapons and gases, and confer an advantage. A few years later the German rocket programme was an early, and influential, example of what the historian Michael Neufeld called 'state mobilization of massive engineering and scientific resources for the forced invention of a radical, new military technology'.[13] It led to the V2s raining down on southern England in the closing months of the war (though, ironically, twice as many people were killed making the 6,000 or so V2s as were killed by them). Something similar might have happened in bio-warfare but as became apparent after the war, Adolf Hitler deeply disliked biological weapons and rejected all proposals to develop them.[14]

The war accelerated the development not just of rockets but also of radar, jet engines and the atom bomb. By then big science in big firms was even more ready to be coopted for war, and the frontiers of science were intertwined with the frontiers of conflict, as when work on anti-aircraft guns, and the tasks of decoding encrypted messages, helped to nurture cybernetics and computing.

From then on, the USA aimed to outspend all its major competitors combined, a goal that was roughly maintained until the twenty-first century. It was an irony of history that it lost most of its wars, despite this superiority, including Vietnam and Afghanistan (which cost around two trillion dollars), with Iraq hardly a glowing victory. However, it didn't lose any wars against comparable nations, all of which wisely steered away from direct conflict.

Many of the frontiers of war are now similar to the frontiers of commercial technology; for example, in grappling with how best to combine human and machine capability. In the 1930s German soldiers such as the tank commander Heinz Guderian sought to combine human guile, courage and skill with radio communications and tanks. Today's armies try to combine the skill of the pilot or drone operator with that of their hardware, with the 'China Brain Project' an ambitious example focused on dynamically combining human cognitive capabilities and state-of-the-art technology;[15] a step perhaps on a path towards wars that humans only observe.

3.4 Science for commerce and growth

The next common goal, the motive for states to support science in this period, subordinated science to the needs of the economy. Nineteenth-century Germany was the paradigm case, at its core a Prussian ideology that sought to scientize the economy as the means to pre-eminence in chemicals, machine tools, cars and later planes.

Britain had a different perspective, celebrating the glories of science and engineering in the Great Exhibition in 1851, but slow to institutionalize science's contribution, a theme that continues to the present day. The US used military spending to spearhead industries, from aerospace to computing, and provided the models for the late twentieth century: venture capital, start-ups, science spawning technology transfers out of universities. It also depended on great corporate centres for research, like Bell Labs, inventor of the semiconductor, fibre optics and more, which was spending around $2bn a year in current money at its peak and employed 25,000 staff including over 3,000 PhDs.

The US has also been a pioneer of state-led commercial development. A recent example is NASA's Commercial Orbital Transport Services (COTs) programme in the 2000s, procuring space services in order to kickstart a space industry that by the late 2010s was booming with commercial launches and satellites, a deliberate echo of the federal government's promotion of a commercial airlines industry in the 1920s through procuring air mail services.[16]

Japan and France had technopolises as well as their own corporate titans, Germany a lattice of intermediary institutions, including Max Planck and Fraunhofer Institutes, alongside firms like Siemens and BASF. Finland had Tekes and SITRA, state agencies to speed up and heat up the transfer of science into business, and thence into wealth.

All of them aimed to be on the frontiers, but at any one point there are surprisingly few places that can genuinely claim to be on the leading edge. A map of the US showing where the most patents come from resembles a flat plain with the occasional sudden spike. Many of the most influential technologies of the last century came from a handful of centres, like

Du Pont or Bell Labs in the 1950s and 1960s or, later, Xerox PARC, Stanford, MIT, Genentech, Monsanto and DARPA. Today the frontiers are no less concentrated, with often a handful of firms dominating a field. ASML in the Netherlands for example, dominates chip manufacture through its extreme ultraviolet lithography. The frontiers of AI may be becoming even more concentrated: 70 per cent of AI PhDs went to work for companies, compared to 21 per cent two decades earlier, while, because of the huge computing power and data needed to train models, companies' share of the largest models rose from 11 per cent in 2010 to 96 per cent in 2021.[17]

Many countries have tried to replicate the frontiers of their time. At one point it seemed that almost every nation had a plan to nurture a Silicon Valley – or rather a Silicon Wadi, Glen or Fjord. A similar pattern can be found in the life sciences, boosted by the extraordinary hopes invested in genomics, the breakthroughs of mRNA technology and the vast potential of synthetic biology. But, despite some trends to isomorphism, the patterns have been varied, perhaps because it is in practice so hard to copy successfully. In some countries a few big firms are dominant, while in others the state leads; in some, globally oriented universities act as magnets, while in a majority of countries the only option is to rely on foreign sources of money to grow capacity. Generic prescriptions for industrial and technology policy remain popular – from aiming to replicate Silicon Valley's venture capital model or reviving 'mission-oriented' innovation – but they tend to fall apart as they come close to implementation: understanding the differences and details of context turns out to matter more than the ability to follow fashion.

Economists have continued to promise that more R&D will in time lead to more growth. A succession of OECD reports and academic treatises suggested an almost linear relationship. But the means of transmission are less clear, as is how much the transmission happens through ideas, how much through people and how much through capabilities. One recent survey summed up the uncertainty: 'Technological progress plays a central role in theories of economic growth. Because social returns to research and development may be larger than private returns, firms may underinvest in innovation, thus reducing the rate of technological progress. Can government-funded R&D fill this gap and generate long-term growth? Despite the fact that governments expend significant resources on R&D every year – over $158 billion in the OECD in 2020 – the answer remains unclear.'[18]

Cherry Murray[19] argues, for example, that although 85 per cent of the US's economic growth derives from research and development, precisely which R&D achieves the impact is less clear, with many high profile and very expensive programmes delivering meagre results, from nuclear fusion and graphene to the $100 billion War on Cancer. NASA received over 0.7 per cent of GDP annually in the mid 1960s and employed, directly and indirectly, over 400,000 workers at the peak of the Space Race. But it's hard to prove, and hardly plausible, that bigger impacts could not have been achieved by spending the money in different ways.

Other widely held assumptions are equally vulnerable to sceptical analysis. It is often assumed that the ideal way to make more economically valuable knowledge is to concentrate money on the most academically excellent institutions, groups and individuals, using competition, peer review and measurements of academic prowess. But the evidence for this claim is thin and, in other periods, countries have prioritized training people as a better route to impact (for example PhDs spreading out into industry). Moreover, the relationship of policy to profit remains problematic. Technological advance rests on public funding for universities and basic research and, in recent decades, on often generous subsidies for start-ups and venture capital. Yet the beneficiaries make windfall gains and only repay through general taxation, which in the case of some industries, such as digital, can be kept very low (which is why so many of the richest people on the planet have made their fortunes in digital, have paid relatively little in tax and contributed nothing to breakthroughs in the underlying technologies on which their wealth rests).

3.5 Science for glory

Both governments and scientists underplay the role of glory, reputation and prestige in science. They prefer to claim more utilitarian goals. But states are creatures of glory and it's hard to understand the actual choices of states through too narrowly instrumental and 'realist' a lens.

This search for glory can be traced back to the pyramids of ancient Egypt, via elaborate technologies used for entertainments: extraordinary displays of water, fireworks, in which rulers could bask in their wealth, impressing their peers with works close to magic. But glory isn't just a feature of the pre-modern world, as Giorgio Agamben shows in his attempts to reconstruct the relationship between glory and economy, king and government. It may be even more important in an era of mediatized, theatrical politics which has brought 'a new and unheard-of concentration, multiplication, and dissemination of the function of glory as the centre of the political system. What was confined to the spheres of liturgy and ceremonials has become concentrated in the media and, at the same time, through them it spreads and penetrates at each moment into every area of society, both public and private. Contemporary democracy is a democracy that is entirely founded upon glory, that is, on the efficacy of acclamation, multiplied and disseminated by the media beyond all imagination.'[20]

The investments in NASA and moon landings are a classic case, much better understood as tools for glory despite some retrospective justifications in terms of spin-offs. Space continues to be the playground for glory for states, from the United Arab Emirates, which sent a probe to Mars in 2021, to India, which has a long history of launches and satellites and has sent unmanned vehicles to the Moon and Mars. It also has a particular appeal for billionaires. It is perhaps better that glory is sought in space, or in sports, than on battlefields, and many citizens enjoy the vicarious thrill of shared glory. But such ventures may be a symptom of how much states are captured. Mature democracies might be more sceptical of expensive quests for glory, their citizens preferring to spend a few percentage of GDP on their children's education than a firework with little to show for it a decade later.

3.6 Science for power over people

States can use science to help them dominate, oppress and manipulate. The science of codes, poisons and tortures goes back millennia. In more recent times the sciences of measurement were primarily developed to help states tax and control a public who had few rights, pioneered by figures like William Petty who introduced rationality and mathematics into the work of government in ways that have now become commonplace. William Petty was a colleague of Hobbes and became a large landowner in Ireland. His 'invisible college' of political arithmetic aimed to ensure that statistics (a word derived from the German 'Staat') would make more rational government possible. He claimed a cool, objective approach, expressing himself 'in terms of number, weight and measure; to use only arguments of sense and to consider only such causes as have visible foundations in nature, leaving those that depend on mutable minds, opinions, appetites and passions . . .'[21] In reality he had an uncanny ability to recommend new laws that happened to benefit his personal wealth.

In France the pioneer was François Quesnay, who promoted 'the idea that if men were to govern according to the rules of evidence, it would be things themselves rather than men that govern'.[22] Quetelet worked on 'social physics'. Others worked on new techniques for accounting, raising taxes, and for moving and processing information. These new techniques treated people as categories and they made states ever more impersonal, and, literally, calculating, in their concern for probabilities and distributions. But their influence was vast: as Ian Hacking put it, there can be no doubt that 'quiet statisticians have changed our world – not by discovering new facts or technical developments but by changing the ways that we reason, experiment and form opinions about it'.[23]

In some accounts, such as James Scott's work, these new methods for 'seeing like a state' are portrayed as coercive tools for centralized, predatory elites.[24] They often were. But the pursuit of knowledge could also be imbued with a mission of moral betterment. In the early eighteenth century, Prussia established the principle that officials in the army, tax system, schools and the postal service had a duty to the people rather than to the feudal nobility. The 'cameralists' who thrived under Frederick William I and Frederick the Great sought the happiness of the ruler and his subjects through rules and laws, and through 'meritocracy rather than noble birth, administrative science rather than feudal law, standardized principles rather than local particularity, and formalism and professionalism rather than traditionalism'.[25]

In Britain, the word statistics was first used by John Sinclair in the 1790s: his massive 'Statistical Account of Scotland' promised to find the 'quantum of happiness' rather than being used solely for taxation and war.[26] Jeremy Bentham deepened the theory with a utilitarianism that promised the greatest happiness of the greatest number, and the application of scientific reasoning to everything from the design of prisons to international trade.

Two centuries later Michel Foucault presented science and social science as tools for managing populations, their health, mental health and compliance. In this perspective knowledge was always tarnished by its connections to power, forcing changes in behaviour and culture to align the people with the interests of the state. Such a lens makes sense of an era in which states use ever more sophisticated methods of manipulation with nudges, behavioural science, the use of game theory, operational research, systems engineering, systems dynamics, and social credit systems.

These ever more complex systems of government, and what Foucault called 'governmentality', became the cousins of popular sovereignty. Such accretions of bureaucracy have had few advocates among people of letters. But while it is fashionable to see these through a lens of suspicion (as Foucault did), that instinctive hostility to organized power sits uneasily with the results: the most profound advances in health, life expectancy and living standards in human history. More controlled societies could also be happier ones, freer from the fear of random violence or the random death of a child.

But how to weigh up the merits of control and standardization against the merits of messy freedom? Writers like Foucault always avoided saying whether they thought these processes were on balance good or bad, let alone what they might have done differently, preferring to stay in a safer space of critical detachment. Today we grapple with similar dilemmas: how far is it acceptable for states to deliberately manipulate and influence their citizens – for example, to gather data to segment populations and send them distinct messages in order to secure appropriate behaviours, such as sending children to school or helping with homework, encouraging vaccination or home insulation? How much do the ends justify the means? Is a scientific state a coercive big brother or a benign guardian protecting us from risks from which we cannot easily protect ourselves? These are some of the questions I turn to next.

4

Science bites back

Protection from risk is the foundation of a state's legitimacy and always has been. States promise to protect us from the risks we can't handle alone. These risks always included the threats of violence and war and, more recently, they came to include the risks of poverty, infectious disease, natural disasters and increasingly the risks that result from science and technology too.[1] Because these risks are beyond the powers of individuals, families and communities, we need states, which is why various families of political philosophy, including anarchism and libertarianism, never managed to build on their undoubted intellectual and cultural appeal to convince whole societies that they might actually work.

Reducing risk is a natural human wish and it is also one of the motivations for investing in science and technology. If we can predict a seismic irruption, we can strengthen our homes or evacuate a town. If we can spot an epidemic before it has spread too far we can shut down transport and social mixing, and speed up work on a vaccine. If we can predict that a particular person has a heightened risk of breast cancer or schizophrenia we can try to reduce that risk.

Economists have suggested that as societies become richer they tend to choose safety over more consumption, or at least lower growth in consumption,[2] and by most measures our lives are far less risky than those of our ancestors. Our chances of being murdered, or killed by an infectious disease are dramatically lower, and there is a good case for seeing risk management in much of the world as a remarkable story of success. For example, the death toll for fatal occupational accidents in Germany fell from almost 5,000 in 1960 to less than 400 in 2014, the number of traffic accidents from 22,000 in 1972 to 3,700 in 2014, and the number of fatal heart attacks and strokes from 109 cases per 100,000 to 62 between 1992 and 2002.[3] Or, to take a very different example, cyclone-related deaths in Bangladesh have fallen a hundred-fold since 1970, thanks to sophisticated warning systems and support for safe return, so that people don't stay with their livestock when the disaster strikes.

But, despite these and many other successes, the relationship between risk and knowledge has become more, not less, complex over time, and this poses challenges to states which wish to appear in control of things. The pioneers of the enlightenment assumed that new knowledge would be an unambiguous good. But eating from the tree of knowledge makes life more complicated, and more morally anxious. It used to be said that if you gave a man a fish he would eat for day, but if you taught him to fish, he would never go hungry. Today it is said that if you teach a man to fish, he will overfish. Knowledge is not always good. Ignorance may not be bliss, but it is at least simpler.

For a century or more, it has been obvious that new science produces new risks: the risk of electrocution from household electricity; nuclear leaks or wars; toxic pollution; novel pathogens; or the risk of being run over by a car or killed in a plane crash. Modern science is implicated either directly or indirectly in most of the risks we face. This has cast the relationship between states and science in a new light. In the early twentieth century, governments had to introduce complex rules to govern the new technologies: the car was accompanied by road markings, speed limits, driving tests and later drink-driving rules, speed bumps, emission limits and more; electricity was accompanied by an equally complex set of rules to protect safety.

By the mid twentieth century, science had become an ambiguous partner for the state. Nuclear weapons could win wars but also bite back: in a nuclear winter there would be no winners. Pollution was visible and shocking: threatening a silent spring and poisoned children. Computers came to be seen as the enemies of liberty and jobs, not just as useful cognitive workhorses.

Governments had no choice but to innovate new methods of governance, most visibly in relation to war and the nuclear threat. Almost as soon as the atomic bomb had been used against Japan, the dangers were so apparent that the USA tried to change the rules of the game, aware of the potential insanity of an arms race. The 1925 Geneva Protocol had banned the use of both chemical and bacteriological weapons and provided a model (though the US had not ratified it). The Baruch Plan, proposed by the US just after the Second World War, suggested eliminating existing stockpiles of atomic bombs after a system of international control was established, with a commission whose decisions couldn't be vetoed by the Security Council. It coincided with the creation of the Atomic Energy Commission to help in the control of nuclear energy and reduce the volume of weapons. But it failed: the features that appealed to the US (that it would be run through a United Nations that, at the time, the US dominated) made it less attractive to the USSR and others.

Scientists offered their own answer. The Pugwash group was founded by Joseph Rotblat in the late 1950s. Rotblat had worked as a divisional head at Los Alamos on the atomic bomb but left when he believed the threat of the Nazis getting there first was no longer plausible. He struggled with the exultation of his fellow engineers in the Manhattan Project (in Oppenheimer's words, making a bomb was a 'sweet' problem). Rotblat had asked what the 'minimum strength' of the nuclear device might be: he was accused of disruption, even disloyalty for his question. Instead, he switched from being a trailblazer of new armaments to becoming a prophet of disarmament, campaigning for moratoriums and for what became the Partial Test Ban Treaty in 1963.

Biological weapons provide another example of collaboration between troubled scientists and equally troubled governments. Jean Guilleman asks in his history of their development 'how do scientists, who are educated to help humanity, justify the use of their privileged knowledge for the explicit goal of killing civilians *en masse*?'[4] To which the answer was that many tried to constrain and lock up the knowledge they themselves had created, campaigning for what became the Biological Weapons Convention in 1975 (now signed by 185 nations), which banned the development, production, acquisition, transfer, stockpiling and use of biological weapons (I look at its strengths and weaknesses in more detail in Chapter 12).

But the frontiers of warfare now include many methods that are wholly unregulated or governed, such as weaponizing geo-engineering technologies to cause droughts, flooding and other natural disasters, means of attack for which there are no counter measures or defences. There are cyberweapons to disrupt and confuse the enemy; prosthetics and enhancements for soldiers themselves, robots, swarms, drones; space wars; and a near future where artificial intelligence could determine the outcome of a war in minutes rather than years.

In all of these examples, science tries to serve the state, and state interests. But it does so in ways that rebound, as the advances prove leaky, prone to being copied, stolen or adapted by the enemy, so paradoxically reducing the security of the nation.

Here we see a more general paradox of knowledge and a defining feature of this new age of risk. Although each step forward in science appears to bring order, with difficult issues settled, and apparently definitive conclusions reached, this is never quite the case. Instead, new knowledge also destabilizes. The new knowledge may be used by enemies and competitors against you. The new knowledge may lead to new risks that hadn't been imagined. It may open up new questions in the minds of people. And the very work of science can itself generate risks, as when a 1979 anthrax outbreak in Sverdlovsk, Soviet Union, resulted from anthrax spores being accidentally released from a secret military facility or when one of the most serious foot-and-mouth outbreaks in the UK resulted from a leak from the Pirbright laboratory in Surrey in 2007.

4.1 We (partly) choose what we fear

The complexities of risk are amplified even further once we take politics into account. Social science has shown that what we fear – and how much we fear it – is in part a matter of choice. Our ancestors were terrified of going to hell or being cursed. Many of us are more worried about being given cancer by a fizzy drink or the whole world going up in smoke because of climate change. Research shows that we exaggerate fears of things that are dramatic, immediate and easy to visualize, and where we don't have any control – like terrorism or air crashes.[5] But we generally underestimate risks that are slow and invisible (like global warming) or where we think we are in control, like driving (in the US about one in a hundred people is likely to die in a car crash; even more now die of opioid overdoses).[6]

The great anthropologist Mary Douglas showed half a century ago[7] – and counter-intuitively – that much of our perception of risk is socially determined. How we see things like nuclear power or GM crops reflects our broader world views – a web of beliefs about hierarchy, individual control, or the authority of experts – as much as objective facts in the world. This has been very visible in risk perceptions of vaccines, and explains why just giving people the facts can have little impact.[8]

Within these patterns there is also another, stark pattern. One of the striking findings of recent work on authoritarian politics,[9] and the motivations of followers of figures like Donald Trump or Matteo Salvini, is that they hate complexity and distrust novelty of all kinds. Ambiguous messages that are open about uncertainty simply don't work well for these groups.

The implication is that risk assessment is unavoidably political and unavoidably judgemental. It is also inherently difficult in the aggregate. What makes sense for one group will not make sense for another – which is why when anti-science messages spread through social media that are targeted, and then responded to with broadcast public information messages that are not targeted, the latter have less impact.

So how do states cope with the tension between their desire to demonstrate that they are knowledgeable, and in control, and the truth that they may be as uncertain as the rest of us? Leaders often believe that the public want them to be omniscient.[10] They are expected to have a view on everything. To admit ignorance too often makes us question their qualifications to lead. Indeed, an unfortunate feature of the modern world is that the most confident leaders are often the most trusted, even if that confidence is founded on ignorance – especially if they control mass media that can reinforce their aura of wisdom.

But how should they, or anyone, handle the risks associated with science? Should they admit what isn't known, or what is ambiguous? And what kind of institutions are not only best placed to assess risk but also best placed to explain it, and the judgements that result?

This is a particular problem for autocrats for precisely the reasons just described. It proves their ignorance. A catastrophe under their leadership can fatally undermine their credibility. So, their first instinct is to cover it up. The USSR's leaders tried to cover up the Chernobyl disaster but couldn't (and were gone three years later). The Chinese leadership at first tried to cover up the emergence of COVID-19 and two years later lost significant authority when they shifted policy to end draconian lockdowns under public pressure.

But democracies also struggle with risk, particularly when public inquiries ask why more wasn't done to prepare or prevent (and there's always someone who warned but was ignored).[11] Even harder are the challenges of admitting methods for managing risk, like putting a cost on human lives, which is unavoidable for anyone planning a transport system for example, and trying to balance costs against benefits, but appears monstrous to the public.[12]

Bureaucracies seek to turn uncertainty into risk, the obscure into the calculable, and this is part of their *raison d'être*. Bureaucratic processes can certainly reduce risk – with inspections, audits and scans. There are impressive examples of institutionalized procedures for analysing uncertainty – like the Netherlands Environmental Assessment Agency, which developed 'Guidance for Uncertainty Assessment and Communication'[13] using steps and checklists to ensure clarity and rigour in understanding the limits of knowledge.[14]

A first step is simply to track what is happening. Many governments keep registers of hazardous materials or algorithms (Singapore has attempted to create an 'Automated Decision-Making Register'). Others monitor the warning signs of floods and hurricanes, epidemics, oil spills and cyber-collapses, which may point to the need for bans or precautionary measures.

But, although science can map and measure, it has no tools for calibrating – how to balance the risks of dangerous artificial intelligence against their potential uses; how to balance the risk of pollution from chemical plants against the potential benefits of those chemicals.

As a result, the practice of risk policy is rarely very rational and can sometimes be absurd. As Gerd Gigerenzer points out, the total deaths from BSE in Europe, during a period when tens of millions of cattle were slaughtered at a cost of tens of billions, were roughly equal to deaths from drinking lamp-oils during the same period.[15] The reason is that although there is plenty of science in risk assessment, there is no science of risk. I learned this first hand when trying to design a risk management strategy for the UK in the early 2000s. In the year 2000 a strike by fuel drivers had almost brought the UK to a standstill. It turned out that in the age of just-in-time production stocks of fuel would barely last a couple of days. Food deliveries and hospitals looked set to grind to a halt, though luckily the strike was quickly settled. Then, less than a year later, the UK faced a severe outbreak of foot and mouth disease. Six million cows and sheep were slaughtered to stop the disease. Both crises showed up the weaknesses of governments' systems for handling risk. I was then running the UK government Strategy Unit and we were given the task of rethinking how risks should be spotted and managed.

Much of government – like most other organizations – tends to focus on the present and tends to ignore potentially harmful risks, particularly low probability but high impact ones. So, we recommended comprehensive machineries within departments and the centre of government to better understand and prepare, looking out for everything from pandemics to financial crises, attacks on critical infrastructure to extreme weather events. We used simulations as well as scenarios and models, to put senior decision-makers, including politicians, through emotionally compelling simulations where they would act out how they might respond in the heat of a crisis (Singapore did this particularly well). Purely paper-based exercises often missed the vital dynamics. A central team (the Civil Contingencies Unit) was set up, and closely tied into local government, so that a network could be quickly mobilized in a crisis. Some of the work focused on finance – including bigger allocations to deal with growing threats, such as increased flooding. And in everything the aim was to find *proportionate* responses to uncertain risks – so that ministers wouldn't be punished for taking risks seriously (like the French minister who spent heavily in response to the threat of the H1N1 swine flu virus in 2009 and was then punished for wasteful overreaction).

One main conclusion was that you could never predict exactly which threats would hit, or the shape they would take. Rather than attempt perfect prediction, it was better to emphasize resilience and adaptability, so that when the crises happened you could respond fast and flexibly. Again, though, that was easier to state as a principle than it was to implement: how much, for example, should any nation spend on health security, or on ensuring it has a manufacturing capability for advanced vaccines ahead of a future pandemic? How much should it spend on ensuring resilience in its communication networks? There are no straightforward answers and those who work on risk face the paradox that their successes are invisible and their failures are very visible, so that they are often either seen as Cassandras, warning of disasters that don't strike,[16] or as myopic, failing to see disasters in advance.

This is a space, though, where taking risk seriously challenges some economic orthodoxies, including the emphasis on efficiency and optimization. To be resilient against risks requires some spare capacity or redundancy in the system. Yet the dominant ideas of the 1980s and 1990s were suspicious of anything that appeared to diminish current efficiency. This led to interesting arguments with the then regulators of sectors like energy, who resisted the idea that they should regulate key infrastructures with a view to withstanding high impact, if low probability events.

These mechanisms for handling risk worked fairly well over the next ten to fifteen years.[17] Similar approaches are now being developed around AI, for example combining registers, assurance and audit, as well as formal regulatory powers. Health security and bio-risk is another field gaining new attention as governments try to prepare for future pandemics and avoid the error of assuming that the next one will be similar to the last.

Rigorous analysis of risk can sometimes throw up surprising patterns. A good example is the risk of blood clots from the AstraZeneca COVID-19 vaccine, which were both exaggerated on social media and then discounted by many experts. Yet the detailed analysis showed a striking distribution of risk. For every 100,000 women in their twenties who were vaccinated there were on average 0.8 ICU admissions prevented and 1.7 blood clots; for women in their sixties, 8.7 ICU visits prevented and on average 0.5 blood clots: in short, a very different balance of benefit and risk.

But the lesson of history is that it is always inherently hard to fully grasp risk, or institutionalize responses. Many technologies that looked promising turned out to be problematic or disastrous, including leaded petrol, chlorofluorocarbons and DDT and, for many others, the jury is still out.[18] This unpredictability became apparent in the examples of the companies we looked at as exemplars when trying to frame a government strategy for risk. In the early 2000s BP, the global oil company, prided itself on its management of risk and saw itself as a model. Yet it was almost destroyed by the DeepWater Horizon disaster in 2010, for which it pleaded guilty to multiple counts of manslaughter and incurred costs of $65 billion, including multi-billion-dollar fines. A decade later, the very countries considered by the Global Health Security Index in 2019 to be best prepared for pandemics – the US and the UK – handled it particularly poorly.[19]

4.2 Predicting global and existential risks

It's even harder to handle the really big risks. The apocalyptic forecasts of Paul Ehrlich in the late 1960s and the Club of Rome in 1972 are good case studies (the latter updated in very interesting ways half a century later).[20] They were right in their broad warnings about the physical limits to growth and the risks of systems collapse. But, in retrospect, they were wrong about population as the key aspect of this and wrong about 'peak oil'. Indeed, the wicked problems of the time – including population growth and endemic inflation – were solved more by politics and social organization than by technology (though fertilizers, new agricultural methods and contraceptives all played their part). Could they have known? And what should have been done differently?

The same questions arise in fields such as biosecurity. The idea that bioterrorism might be a serious threat took off in the 1990s, though it came up in fiction and scenarios long before, describing lab accidents, malign use of gene-editing,[21] and pandemics, whether natural or deliberate. Mass travel makes it easier for diseases to spread. Encroaching on animal habitats speeds up transzoic spread, further amplified by war and the displacement of peoples. The risks are scary, beyond comprehension. They look dynamic and potentially exponential and so occupy a vivid place in popular imagination. Likewise, many terrifying predictions have been made about widespread access to gene-editing technologies.[22] A recent experiment used machine learning to devise new drugs, but flipped the methods to see if dangerous organisms could be found. The answer was that many could be found very quickly, feeding profound anxiety about just how easily science could run out of control.

Yet, so far, the worst fears have not been borne out and many experts view the risks as manageable. Governments could have spent huge sums preparing to respond to acts of bioterrorism – but in retrospect this would have been wasted. So what should they do? How should they calibrate the scale and severity of their response? The dilemmas were apparent in 2011 when US and Dutch scientists researched how a modified H5N1 bird flu virus could become transmissible between mammals. When officials looked at the security implications they warned against publication.[23] 'It is the nature of the influenza virus to cause pandemics. There have been at least eleven in the last 300 years, and there will certainly be another one, and one after that, and another after that.'[24] Here at least was a clearcut decision that didn't cost much.

We might hope that we could be scientific about risk, quantifying it, analysing it and bringing objectivity to bear. But, as Roger Pielke and others showed in relation to hurricane risk and insurance, this is much harder than it appears, with models offering 'truthiness' rather than truth even in such apparently straightforward cases.[25] Other cases are more ambiguous. I once had to advise on regulations around mobile phones, when many were convinced that if they were held next to our ears for several hours a day they were somehow likely to threaten the health of brains, and in particular to cause brain cancers. Given that phones emit radiation, with 2G, 3G, 4G emitting radiofrequency in the frequency range of 0.7–2.7 GHz and 5G up to 80 GHz, and given that phones do heat up the ear and brain, this was not an unreasonable worry, even if they are low frequency, low energy and non-ionizing.

Science suggested that the risks were unlikely, but no one could be definitive. In one study that followed more than 420,000 cellphone users over a twenty-year period, researchers found no evidence of a link between cellphones and brain tumours, and many other studies confirmed their safety. But the key point here is that the studies covered periods in which billions of people used phones, and at an earlier stage the fact that there was no evidence of an effect was not evidence of no effect.[26]

The legitimacy of scientific discourse on risk is also itself risky. The experts have to warn, but if they warn too much (and the risks don't materialize) they lose credibility. If they disagree, this may call their expertise into question. Brian Wynne argued that once you see that 'scientific expert knowledge . . . embodies hermeneutic (and formulaic) and not only propositional truths',[27] it's easier to question and doubt it. And since how people make sense of risk depends in part on their own experiences and observations, the public can easily call into question experts who appear *parti-pris* and overly tied to particular interests, as so many are.[28]

As Mary Douglas had put it: 'selective attention to risk, and preferences among different types of risk taking (or avoiding), correspond to *cultural biases* – that is, to worldviews or ideologies entailing deeply held values and beliefs defending different patterns of social relations'.[29]

Selective attention also reflects what we see. In the 1980s the US Environmental Protection Agency undertook a large project to compare the views of the public and its own experts on risk and found great divergence. The public were worried about nuclear accidents, pollution and hazardous waste, which were prominent in the media, yet the experts didn't see any of these as priorities. The experts thought that indoor air pollution and radon were particularly important. It's appealing to believe that the public are generally wise in their views. Yet the experts will generally have spent far more time thinking about the questions and will have more facts at their disposal. The only plausible conclusion is that we need to attend both to the detached analysis of risk and the perceptions, taking both seriously.

These perceptions are then bound to vary according to context. For example, one recent survey of attitudes to policies on climate change showed that in China, the UK and the USA, positive frames increase the likelihood of public support for action, while negative frames reduce the likelihood of public support. Elsewhere the opposite is true.[30]

So, an era when science is bound up with risk needs different methods and different mindsets. It needs recognition that the objective and the subjective, the scientific and social, are intertwined. How we see risks depends on what we think matters. Moreover, an era so shaped by risk requires humility.

The oddity of knowledge is that it also opens up a new space of ignorance. The more we know, the more we know we don't know. For example, 80 per cent of matter in the universe is thought to be dark matter, but this remains just a conjecture. There are glaring gaps in medical knowledge – for example around the workings of immune systems. Strange patterns emerge – for example, new patterns of cognition in computational 'large language models' – that no one quite understands. As W. Ross Ashby wrote in the 1950s, physics and chemistry 'gained their triumphs chiefly against systems whose parts are homogeneous and only interacting slightly'. Faced with more complex challenges, we 'must beware of accepting their strategies as universally valid'. Artificial intelligence in all its forms exemplifies the challenge. Governments are trying to bring in new rules to handle possible risks, from Europe's AI Act, which seeks to ban algorithms that exploit vulnerabilities, use social scoring or do biometric identification at a distance. China has introduced strict controls too, while the US has proposed an AI Bill of Rights. But, all recognize that the pace of change means that any specific proposal will soon be out of date, and even any principles may not prove durable.

Instead, it may be wiser to create institutions with the freedom to adapt and reinterpret their roles, which can change as the facts change, rather than having their roles specified too precisely in law.

Any such institutions, however, face a bigger challenge; a challenge that faces anyone seeking to design public organizations. Ideally, the shape of an institution should fit the shape of the task or problem with which it is dealing. In the case of many risks, particularly low probability but high-impact ones, these require a longer timescale than that of governments. In other words we need institutions that are at least partly insulated from politics, an issue I look at in more detail in Chapter 10. And, since many risks don't respect national boundaries, we also, ideally, need institutions that are supranational, a topic I look at in Chapter 12.

For now, though, let me just re-emphasize the importance of humility. The world remains vastly more complex than our minds or our representations, which is why no one can make confident predictions about issues such as bio-risk or artificial intelligence. The map is not the territory but just a map and we see the world through a glass darkly.[31] As Alfred North Whitehead put it, 'Not ignorance, but the ignorance of ignorance, is the death of knowledge.'

In these cases of profound uncertainty the best we can hope for is to map out potential scenarios of risk; to identify potential sequences or early warning signs; and to ensure that there are observatories scanning for these, that can warn those who have the power to mobilize action, which has to come from politics.[32]

5

The scientist's view of politics as corruptor

Over the last decade I have heard dozens of entrepreneurs and computer scientists argue that it would be outrageous for governments to constrain them and others working on the frontiers of artificial intelligence. Perhaps they might do so in a decade or two, but not now. Regulation and bureaucracy in their minds are enemies of genius and enemies of scientific imagination, and anyway far too slow and lumbering to keep up with the pace of technological change.

Although, very belatedly, in the mid 2020s scientists began to recognize that regulation might, after all, be necessary, their antipathy to government policies and rules fits well in a long tradition of scientific argument making the case for self-sufficiency and a minimum of interference. That view can be traced to the 1660s, when the Royal Society started to publish *Philosophical Transactions*. This was arguably the first scientific journal and creator of the procedure of 'peer review' whereby research and ideas are critiqued and improved by a community. This moment is sometimes seen as marking the birth of a modern kind of science, based in its own institutions, clear about its ethos and committed to 'the establishment of science as a distinct activity, to be controlled only by its own norms'.[1]

I mentioned earlier that the norms and values were summed up in the slogan *nullius in verba*, meaning we should believe no one's word, and take nothing at face value. Ever since this moment, scientists have become more confident in asserting a distinctive ethos and philosophy. Theirs' is a world of facts, theories, experiments, proofs, embedded in a widening circle of epistemic institutions – including schools and universities as well as laboratories – that promote insight and understanding, a world assumed to be superior to the grubbier world of politics, compromise and half-truths. That didn't always guarantee what we would now see as wisdom: the fact that 'many leading proponents of the Royal Society gave scientific credibility to belief in witches by presenting experiment-based arguments for their existence' is just one of many embarrassments that paved the road to progress.[2]

For some – like H.G. Wells and many thinkers of his time – the hope was that science would displace politics. Scientists often see politicians as fools unable to grasp the basics, in best useful sources of money, at worst ignorant monsters, and politics often gives them confirming information. On 22 January 2020, in the very early stages of the COVID-19 pandemic, President Trump promised: 'It's one person coming in from China, and we have it under control. It's going to be just fine.' Later, he suggested using disinfectant as a treatment. Brazil's President Jair Bolsonaro thought it all fake news. There were plenty of others, like the President of Tanzania, who recommended lemon and ginger as a cure, or Madagascar's President, who developed his own herbal remedy, or Nicaragua's Daniel Ortega, who just saw the pandemic as a message from God. All fared badly.

So, in one view, politics is essentially a throwback, at best irrelevant, at worst disastrous. It is a legacy of earlier periods of ignorance. A good world should make politics subservient to science, using its logics and making the most of its methods, instead of the hubris of ignorant leaders who are not even aware of their ignorance.

Data can encourage this view, since it shows that scientists have remarkable legitimacy and trust. According to one recent international study, 'scientists and their research are widely viewed in a positive light across global publics, and large majorities believe government investments in scientific research yield benefits for society'. In the US, 'confidence in scientists is far higher than in politicians or bureaucrats. The so-called populist backlash against science and expertise . . . is a figment of the imagination, itself in the land of opinion and post-truth.'[3]

As we'll see, the more detailed patterns are not so reassuring or so simple. But they appear to confirm a view of science as not just separate from politics but also above it. These patterns were set quite early. In their study of the origins of modern science, Steven Shapin and Simon Schaffer showed how the pioneering seventeenth-century scientist Robert Boyle established a new political status for science, in part by denying its political character. Through using experiments scientists could establish a firmer foundation for knowledge, so separating the constitution of knowledge from the constitution of power and establishing a world view in which science was both above, and separate from, politics and society. Hobbes, by contrast, thought this a route to division and disorder. Arguments about both kings and things fed each other.

Boyle's perspective became widely accepted, making the scientist duty-bound to pursue knowledge wherever it led, and to resist incursions. As Isabelle Stengers put it, scientists 'actively took part in the ongoing construction of an asymmetric boundary that would protect their autonomy and resist intruders'.[4] This view could be justified in pragmatic terms. Thomas Kuhn, for instance, argued that the scientist's keen interest in puzzle-solving for its own sake led to breakthroughs, whereas too much concern for applications and external rewards tended to be bad for scientific progress.[5]

Many strands of liberalism presented this autonomy as both good in itself and as aligned to a liberal world view that was thought to be similarly empirical and objective. In politics, science could therefore act as a neutral honest broker, presenting facts as they were. This value system entailed a distinct set of practices, best described by Robert Merton, who argued that these were organized around four clusters of principle.

The first was 'universalism': no one should be excluded from scientific discussions because of their social characteristics, their class, ethnicity or background. A second was 'disinterestedness': the culture of science should uncover any biases and hidden political agendas, dealing with the facts as they are. The third was 'communalism', which prescribed that scientists should share their method of analysing and collecting empirical objects as well as their results. Finally, there should be 'organized scepticism', with truth claims subject to examination by peers.[6]

The first version of Merton's paper was published in 1942 in the context of the war against fascism and presented science as part of democracy. Indeed, it argued that true science could only thrive in a democracy. Yet the model also required science to resist incursions from politics; to govern its own financial allocations, the valuation of projects and priority setting, using review by peers as the main way to guide decisions (Merton also showed how parts of the scientific ideal drew on religious faith).[7]

In 1945, Michael Polanyi made the case for autonomy in even starker terms, attacking the argument that 'since the contents of science, and the progress of science both vitally concern the community as a whole, it is wrong to allow decisions affecting them to be taken by private individuals. Decisions, such as these, should be reserved for the public authorities who are responsible for the public good. It follows that both the teaching of science and the conduct of research must be controlled by the State.'

This, he argued, was 'fallacious' and wrong. Indeed, 'if attempts to suppress the autonomy of science . . . [succeeded] . . . the result could only be a total destruction of science and of scientific life . . . and the main body of science itself would disintegrate and fall into oblivion'.[8]

At roughly the same time Vannevar Bush successfully argued that such autonomy would in the long run pay off for the state. Free scientists would be productive ones. This argument became conventional wisdom and, in many countries, scientists won a substantial degree of autonomy, in part because of the negative examples of Nazi Germany and the USSR and in part because of the prestige they gained through visible problem-solving. Wars amplified this prestige – not least the astonishing successes of Bletchley Park in the UK, the invention of radar or the Manhattan Project. Disasters both short and long term, from Fukushima to COVID-19, drought and climate change, had a similar effect. During such moments of intense stress, the world view of a sovereign and substantially autonomous science was vindicated, and these were times when science bypassed or trumped politics, essentially because it worked, according better with reality than the alternatives.

5.1 The ideal of self-government

If science really is superior, and more aware of opportunities and risks, then it follows that scientists should indeed make the decisions on who gets funding and how science is used. This argument shaped the creation of research councils, peer review and much more: self-referential systems with their own hierarchies of authority.

Even in Bacon's Atlantis, the House of Solomon is autonomous, free to decide who should know. As Bacon describes the various 'sound-houses', 'perspective houses', dispensaries, diverse mechanical arts, engine houses, and a mathematical house, he emphasizes their autonomy: 'And this we do also: we have consultations, which of the inventions and experiences we have discovered shall be published, and which not; and take all an oath of secrecy for the concealing of those which we think fit to keep secret; though some of those we do reveal sometime to the State, and some not.'

That ideal of self-government inspired many scientists to try to lead not just in creating new knowledge but also in shaping its effects, offering themselves as better interlocutors than mere politicians or bureaucrats. I've already mentioned the work of scientists campaigning for nuclear disarmament and bioweapons treaties and there are many examples of scientists attempting to lead through diplomacy in defiance of the Cold War.[9] Something similar happened in the relatively early days of genomics when, in 1975, the world's leading molecular biologists met in Asilomar, California, and drew up guidelines on which kinds of experiments should not be done. Their aim was to establish some baseline ethics and safety standards to stop human-made genetically modified organisms threatening public safety.

A generation later, after the disastrous first gene-editing of human beings, leading scientists including Emmanuelle Charpentier, Feng Zhang and Eric Lander, called for 'a global moratorium on all clinical uses of human germline editing – that is, changing heritable DNA (in sperm, eggs or embryos) to make genetically modified children' and 'an international framework in which nations, while retaining the right to make their own decisions, voluntarily commit to not approving any use of clinical germline editing unless certain conditions are met'.

A parallel event to Asilomar brought together artificial intelligence pioneers in Puerto Rico in 2015. Here, figures such as Demis Hassabis, the founder of DeepMind, Stuart Russell, Max Tegmark, Jaan Tallinn and Elon Musk came together to discuss what rules might constrain AI. Having discussed the risks of an intelligence explosion they turned to the challenges of AI safety. Their open letter committed to preparing for the risks that AI could bring. It argued that 'because of the great potential of AI, it is important to research how to reap its benefits while avoiding potential pitfalls . . . [through] expanded research aimed at ensuring that increasingly capable AI systems are robust and beneficial: our AI systems must do what we want them to do'.

This was a worthy attempt to match power with responsibility. But, unfortunately, the letter was surprisingly devoid of social science (it brought together technologists and philosophers), as signalled in the unreflective use of the word 'we' in the letter.[10] It was also devoid of politics (reflecting a cultural base that tends to disdain government) and oddly uninterested in discussing what forms of governance might be useful. In retrospect their initiative showed the limits of scientific self-government rather than its power.

The same year I had published proposals for a Machine Intelligence Commission, suggesting how societies could cope with the impacts of AI on so many fields, from finance to aviation, cars to policing. Its role would be to sit above existing regulators and help them to navigate the development of new rules on issues such as facial recognition, often taking an experimental approach. Not long after, the UK government created a Centre for Data Ethics and Innovation, a weaker version of what I had proposed, and the Centre went on to do serious work on issues such as targeted advertising. But there was a disconnect between, on the one hand, such practical work and the public institutional design that lay behind it, and on the other the many gatherings of the AI leaders, which produced impressive rhetoric but little action.

In the years that followed many came to view much of the work on AI ethics as a distraction rather than a solution. Dozens of books and conferences worried about issues such as the moral dilemmas of 'trolly problems' (how should you, or an algorithm, weigh up the lives of children or the old, if your vehicle threatened to crash into them, for example). Others explored the possible dynamics of 'the singularity' a generation or more into the future, when AI might suddenly take off and dramatically supersede human intelligence.

Yet, at the time, AI was being applied in almost every area of daily life with very little serious analysis of the social and ethical issues. Governments struggling to know how to handle the use of algorithms in pensions, flash trading, cars or shopping were given little help by the academy. The work on the distant horizon was intellectually appealing but arguably a diversion from the harder problems of the present.

Serious work on AI ethics is vital. But the cynics were justified in coming to see the extraordinarily generous funding of AI ethics centres as a classic example of misdirection,[11] with methods that tended to ignore the power dynamics of current and future applications; the social distributional effects in terms of winners and losers; the asymmetries of understanding, and much more.

Here we see both the power, and the limits, of what has sometimes been called the 'Republic of Science' ideal, the notion that science and scientists can govern themselves. Interestingly, a parallel initiative to the Puerto Rico gathering happened a few years later, in 2023, and involved many of the same scientists and entrepreneurs (including Elon Musk and Max Tegmark). The open letter they published then recognized the limits of voluntary action: 'we call on all AI labs to immediately pause for at least six months the training of AI systems more powerful than GPT-4. This pause should be public and verifiable and include all key actors. If such a pause cannot be enacted quickly, governments should step in and institute a moratorium.' But there was, once again, nothing on how this might be done. In the US, the public apparently supported such a pause (by a margin of roughly five to one):[12] but without a serious plan there was little chance of it happening.

In some cases, civil society has tried to fill the space between governments and scientists, expressing a public interest that is distinct from the interests of the state. Many fields of applied science have sophisticated systems of self-governance, distant from any formal state power. A British example is Trinity House, which, for half a millennium, took charge of maritime safety in coastal waters, providing technologies, tools and training. It's paid for by fees from commercial vessels and an endowment and is governed by a committee that appoints itself.[13] Another example, based in the US, is the 'Underwriters Laboratory', which for over a century has run sophisticated programmes in safety science, using councils of practitioners and academics to solve practical problems.[14] There are also many examples of independent initiatives to establish norms, like the 2021 'Tianjin Biosecurity Guidelines for Codes of Conduct for Scientists'.

Philanthropy has also played a crucial role in supporting the independence of science.[15] The Wellcome Trust in the UK became one of the world's largest funders of medical research, able to collaborate with even big governments on equal terms, as did the Gates Foundation. The large US foundations had over many decades used their wealth to accelerate science, as when Rockefeller prompted the 'Green revolution' in the 1960s, or the Howard Hughes Medical Institute invested in health (with spending of nearly $1bn a year). Although in the USA the contribution of philanthropy is barely a tenth of that of the federal government, it provides freer money with fewer strings or expectations

But philanthropic power is never wholly pure and it is neither accountable nor neutral – it is inevitably shaped by interests and ideology, and can turn power in one domain – business – into power in other domains, such as healthcare in the case of the Gates Foundation. The huge funds raised by the Effective Altruism movement gave them the power to fund so many of the researchers in fields such as bio-risks that few dared criticize them. Yet, so long as the relative scale of philanthropy remains limited, providing a diversity of sources of funding rather than a new monopoly, it tends to be, on balance, benign.

5.2 The philosophy of autonomy

Implicit in the claim of autonomy is the notion that the pursuit of knowledge is inherently good. Science cannot be the servant of something outside science. Art makes a similar claim to autonomy: it cannot be explained or interpreted with criteria that come from outside art, even though many try.

But can this argument be coherent? The idea of a sphere of activity wholly detached from uses, values and interests is hard to sustain. If humanity disappeared, would this activity still be plausible, desirable or wise? If the knowledge led to the destruction of things dearly held, values, institutions and, at the extreme, people, would this be problematic? From what vantage point could these questions be answered, if not that of living, breathing humans? In a country like India, for example, how should the needs of the 300m or so people who have to defecate outdoors be weighed against the virtues of a space programme – and why should scientists be expected to make such a judgement?

Any consideration of these questions tempers the claims to autonomy and usually leads to a more conditional position, a pragmatic rather than principled argument that autonomy achieves more than subservience. Knowledge wholly detached from context or meaning is hard to justify. Some connection to a community with interests, and one that is located in time and space has to be acknowledged at some point. That community can be wide – humanity imagined over centuries or millennia[16] – or it can be narrow, a nation or city at one moment. But it cannot be limitless. Our intelligence works best with a degree of autonomy from immediate pressures, whether of impulses or interests. But it gains its purpose from a place in the world, from how it comes back down to earth.

As a result, the ideology of autonomy has been made conditional, with an explosion of competing theories in the last few decades challenging the simple model: 'mode 2' science, the triple helix,[17] national systems of innovation,[18] responsible research and innovation, embedded autonomy,[19] and ideas that science policy-makers have championed themselves, such as 'frontier research' and 'grand challenges'.[20]

Scientific autonomy has also become more suspect because of the influence of critics who have shown the extent to which science can reflect the interests and ideologies of dominant groups.[21] A tradition associated with Lewis Mumford, Jacques Ellul and Langdon Winner[22] argued that the very autonomy of science was malign, and liable to be oppressive, the ally of power rather than of the people. Donna Haraway[23] and other feminist thinkers like Emily Martin and Evelyn Fox Keller[24] criticized science's spurious claims to autonomy and universality, again seeing science as embodying patriarchal assumptions (a critique that was given extra force by the overwhelmingly male make-up of the author lists of the various manifestos and open letters mentioned earlier).

The switching views can be seen vividly in relation to digital technologies (from the Internet to blockchain), which were at one point thought likely to promote greater freedom, equality and democracy, but then came to be seen as likely to lead to greater inequality, a corrosion of democracy and the bypassing of public institutions. The Science Technology Studies (STS) field that grew out of these critical traditions provided rich descriptions of science history,[25] which demolished the more complacent pictures of scientific virtue, the lazy assumptions of technological determinism and the genius stories of invention.[26]

The criticisms of the blind spots of decision-makers were often telling, not least in relation to gender. In some cases, too, the STS field made a strong case for more public, and democratic, input to science. It showed how 'lay people in communities disrupted by environmental pollution have marshalled knowledge and data through a process of "popular epidemiology"; how sceptical sheep farmers relied on their everyday experience to challenge experts' abstract models of radiation dissemination that was itself based on incorrect assumptions; how lay activists successfully challenged the conduct of science around clinical trials for life prolonging AIDS drugs; how consensus conferences offer a model for citizen engagement in realms traditionally involving scientifically complex technical decision-making; and how French muscular dystrophy patients organized to finance and contribute to research in this area . . .'[27]

5.3 Self-doubts

While science clings to an ideal of self-sufficiency, it has in practice ceded autonomy to governments and businesses as a route to greater power, prestige and money. In this respect it is not unusual. All social groups seek autonomy, yet their status and power also depend on collaboration with others. For scientists to survive and thrive they have to cope with the reality of their subordination to the needs of the military; governments; big business or civil society, negotiating through the thickets of committees, reviews and commissions that feed into the equally prosaic world of regulations and policies. It is the world within companies of teams and projects, targets and ROIs, technology readiness levels and market research and, in academia, of impact measures and technology transfer. These are the places where scientific autonomy is negotiated, detail by detail. Autonomy is also limited in other ways, since science is, perhaps more than ever, part of its wider culture and scientists, like artists, can no longer use the claim of genius to insulate themselves from demands that they be accountable for their behaviour, whether sexual harassment or contributing to climate change.

This shift, as science has compromised and become more enmeshed into its wider society, has reinforced a growing sense of self-doubt that has accompanied burgeoning power. A century ago, Ludwig Fleck proposed in his book *The Analysis of a Scientific Fact* the idea of science as a community. He showed how facts arise out of social processes rather than simply existing out there waiting to be found. Clusters congregate around particular paradigms, or what were later called 'basins of attraction', models that dominate and attract, configurations of ideas that are stable over time, and which are sustained by communities of scientists.[28]

This sociological view implies that truth is whatever the community believes to be true. Fleck's ideas influenced Thomas Kuhn and led some to conclude that science was only an artefact of community, encouraging many writings on the history or anthropology of science to emphasize its contingency, its political nature, the cultural patterns of discovery and argument, and to dial down the extent to which it could be seen as universal, logical or linear.

In the 1940s, the influential LSE professor Harold Laski had warned that 'expertise . . . sacrifices the insight of common sense so that experts tend to neglect all evidence which does not . . . belong to their own ranks' and, a generation later, scholars tried to map out a more humble stance. Alvin Weinberg, for example, proposed the idea that while science produces some answers it rarely has all the answers. His concept of 'trans-science' described questions of fact that cannot be answered scientifically, such as the probability of extremely improbable events, or questions of individual as opposed to aggregate behaviour, or in engineering, everyday judgements about the behaviour of materials.[29]

Bruno Latour and others took this self-doubt a step further, arguing that scientific facts are produced by scientific practice rather than having an independent reality. To many scientists this was abhorrent, even absurd. In a famous incident, the physicist Alan Sokal challenged Latour, if he really did doubt the existence of an external reality, to prove it by jumping out of his window on the twenty-first floor.

Latour sometimes took positions as much to prompt and provoke. I worked with him on some projects, and he was always engaging and entertaining: I saw how much he enjoyed being ironic and flippant and being taken more seriously than perhaps he had intended. He felt that scientists were too comfortable in the idea of a world governed by common theories and understandings, which was rational, universal and progressive. He believed that his emphasis on the construction of facts did not preclude an external reality: as with the job of constructing a building, the job of constructing facts could not be done only with social relations.

But his arguments gained a momentum of their own, and jumped from the entirely plausible view that science works as a community and constructs its own interpretations and truths, to the view that this is all it is: that the strength of a fact depends only on the work needed to undo the network of associations that created it.

'Give me a laboratory and I'll raise you the world', he once said, yet his more considered position was that facts are supported by institutions: 'facts remain robust only when they are supported by a common culture, by institutions that can be trusted, by a more or less decent public life, by more or less reliable media'.[30] As I will argue later this is as undeniable as the more extreme view, that facts are *only* social, is not.[31]

So, too, is what was sometimes called Thomas's theorem, the idea that if people define situations as real they are real in their consequences. Again, it is entirely feasible to accept this without denying the existence of a reality beyond human observation and interpretation. After all, in engineering, the more extreme accounts of social constructionism never made any headway because reality provides such rapid, and clearcut feedback.

The struggles between what are sometimes called realist positions, which see science as exploring an objective natural world, and the constructivist positions, which see science as providing constructions, linking mental models filtered through our senses or instruments, have dissipated. Although the two positions are hard to reconcile in theory,[32] in practice they have been superseded. Pierre Bourdieu's subtle analyses of the power plays both within science, and around science, helped in this respect, recognizing that although science is a social construction it works in ways that ensure its products are not just social constructions, but correspond with an external reality.[33]

But these often heated arguments have left a series of gaps. They left a deficit of design – remarkably few good proposals for how to govern, shape and guide powerful new fields of technology such as artificial intelligence. The radicals preferred a position of detached critique, some in the spirit of Latour or Foucault, rather than committing themselves to potential solutions. They exemplified the dyna-phobia, or fear of power, that affects many areas of intellectual life. Power is correctly seen as sometimes corrupting and often tarnished with compromise and difficulty, leading many to conclude that it's easier to remain at one remove from it, making elliptical pronouncements but avoiding any position that could be held to account.

In some cases this leads to an excess of diagnosis without prescription: for example, Shoshanna Zuboff's work on surveillance capitalism, which drew on decades of critical work by many academics, provided hundreds of pages of often-fascinating critical comment but not a single proposal as to what should be done. In other cases dynaphobia leads to a quasi-anarchist position, which is suspicious of any new national or global public institutions on the grounds that they might be captured by the dominant powers, or global corporations. Only bottom-up movements are seen as truly legitimate, ignoring the historical lesson that movements only truly move the world when they mutate into institutions. The result has been a deficit of options, of useable proposals for shaping or governing the multitude of new technologies pouring into the world.

This deficit is not helped by the myopia of scientific elites. The French historian-philosopher Ernest Renan said that nations are defined as much by what they choose to forget as what they choose to remember. The same could be said of elites, which are defined by what they choose not to talk about as well as what they do.

Science elites sometimes seem to be bound by a shared myopia, as if it is almost a condition of membership not to talk openly about difficult questions (some of which I address in later chapters): should science and technology be guided by the public's priorities, rather than only those of science itself, or the state or business? Can any society choose different technological pathways or are these fixed or pre-determined? Are there solutions to the declining productivity of science, which connects to its broader social contract, since so much science is either funded or subsidized through tax? These questions are often ignored, pushed out of the frame. Yet they are among the most important questions on the boundary of science and democracy.

Meanwhile, the persistence of the ideal of autonomy has left science and scientists in an uncomfortable position. They have become used to an assumption of innocence and virtue. But now, as their power and influence have grown, they find themselves under attack from figures who are truly anti-science, questioning its methods, its results and its morality, as well as facing criticism from others who wish to apply the sceptical ethos of science to science itself, and so to observe it more rigorously. The result is an unhealthy brittleness. Any questioning, however well intentioned, can be rejected on the grounds that it just provides fuel for anti-science interests (and sometimes it does), which makes intelligent debate harder and arguably impedes progress.

PART III

The Problem of Truths and Logics

6

Master, servant and multiple truths

Humanity has not liberated itself from servitude but by means of servitude.

Hegel

States depend on truth. Truth can reinforce and uphold their power and rulers want the endorsement of arbiters of truth in order to feel legitimate: how can they justify their rule if they have no superior insight, knowledge or wisdom, and if others clearly know more than them? Shared truths also support social order. In the words of Michel Foucault, 'the more [government] pegs its actions to the truth, the less it will have to govern . . . governors and governed will be, as it were, co-actors of a drama they perform in common'.

But the truth is slippery and sometimes threatening. Emperor Justinian closed down the Academy of Athens in the sixth century. The library of Alexandria was not the only one to be put to the torch and a millennium later Galileo was put on trial because of how he challenged the church's proclaimed monopoly of truth.

Heretical knowledge was always challenging for rulers. But rulers could rarely just impose their own truths. Instead, they needed allies, and these providers of different kinds of truth were precursors to the advisers and managers of contemporary science, each with an ambiguous relationship to the powerful, offering both a vital service but also a potential threat.

As we've seen, one prominent group were the engineers, the makers of things, who could be judged by whether their bridges and buildings stood or fell. But there were many other categories whose knowledge was more ambiguous. One of the most ancient is the spy chief, the curator of secret knowledge and master of intelligence, who roots out threats and monitors the population for dissent and reports back to the ruler alone. But such masters of intelligence, who may operate at the frontiers of science with their use of cryptography or surveillance, were and are, always dangerous because they can switch allegiance or juggle multiple allegiances. Rulers have few ways to verify their claims and it's easy to manipulate a ruler into seeing friends as enemies. Yet the knowledge the spies provide is essential for survival.

Another ancient role was the prophet who claimed to speak for God or some other mysterious power, revealing truths about the present or future. Because their claims cannot be verified it is their passion or conviction that matters most, the vividness of their language and imagery as they warn and promise. They have a complex relationship to power, often winning a hearing from kings and emperors, but also distrusted and feared for the potency of their messages. As we shall see, some contemporary scientists see their role in a similar light, warning of existential threats in the face of inertia and incomprehension. Joseph Rotblat, founder of Pugwash, is a classic example and many made a similar switch to him – from insider technician to outsider disruptor. Some took on roles not dissimilar to Old Testament prophets, warning of disasters ahead: a Silent Spring, an imminent population explosion, catastrophic war or, like James Lovelock, suggesting a planet with its own regulatory mechanisms and providing material for a quasi-religious view of the biosphere, connecting into the many social movements that now have a scientific wing.

While the prophet warns and inspires, another role is lower key: the seer who predicts. Astrologers remained influential in the modern era (for figures as varied as Ronald Reagan and Indira Gandhi). Others offer skills in forecasting, from the tyro-mancers who read patterns in cheese to predict the future to today's foresight teams, providing scenarios and models of all kinds to grasp what might lie ahead.

Then there is the role of the wise sage. The sage speaks for him or herself, keeps apart from the world, generally resists the temptations of power and luxury, and is often enigmatic in guidance, like the Delphic Oracle that 'neither says nor hides, but offers signs'. The tone of the sage is opposite to that of the prophet. Where the prophet is loud, insistent and passionate, the sage is quiet, calm and cautious. The sage nudges the leader towards insight rather than spelling it out. Often in the past the sages were blind, perhaps because this helped them avoid being distracted by the superficial appearances of things. They perform their role best if they discomfort (in this respect their role overlaps with that of the jester), but they do so in gentle ways. Occasionally scientists have taken on a similar role, offering wise counsel in secret.

A related category is the truth-teller, the embodiment of the ideal of 'parrhesia', the 'free spoken-ness' which means talking honestly about what is and avoiding the deceptions of flatterers.

Modern scientists combine in different mixtures the traditions of all of these bringers of truths to power: they are sometimes engineers showing ways to fix problems, sometimes spies providing secret knowledge, sometimes seers or prophets providing knowledge of the future, sometimes sages providing wisdom or truth-tellers saying how things truly are. All are essential to the work of governing. Their relevance derives from the fact that sovereignty evaporates when it loses truth. And, although power can disregard truth, or be purely cynical, it exposes itself. Power that is purely naked, without trappings, is weak.

Originally the truths that mattered concerned the cosmos, the climate, hunting grounds or competing tribes (knowledge that reduced the uncertainty of the environment and widened the scope for human sovereignty). Later kings drew legitimacy from their own godlike character and carried out rituals to maintain order in the world and to reinforce religious truths. The Mughal ruler Akbar in sixteenth-century India is sometimes seen as the first great leader to promote reason as the highest value of his state, at a time when Islam had much to teach Christianity about tolerance and enlightenment. Then, during the eighteenth and nineteenth centuries many Western governments legitimized themselves by reference to knowledge and reason, with constitutions founded on truths that were taken 'to be self-evident' (including the preamble to the failed early twenty-first-century European Constitution, which acknowledged the primacy of reason). In this way science became a vital source of legitimacy.

The Soviet constitution described the state as 'guided by the ideas of scientific communism' and India's constitution (in the 1970s) asserted that 'it shall be the duty of every citizen of India to develop the scientific temper, humanism and the spirit of inquiry and reform'. First Prime Minister of India, Jawaharlal Nehru, in his book *The Discovery of India*, had explained that 'scientific temper is a way of life – an individual and social process of thinking and acting which uses a scientific method which may include questioning, observing reality, testing, hypothesizing, analysing and communicating'.

This idea of states infused with and informed by science became very popular in the mid twentieth century, and was promoted both by the USSR and by the US, where dozens of foundations were set up to promote science and social science, championed by figures like Julian Huxley (later director of UNESCO), who admired both the USA's Tennessee Valley Authority as an 'experiment in applied social science' and, in 1931, the Soviet Union as an example of 'the spirit of science introduced into politics and industry' (you can guess he didn't visit Ukraine at the time, where historians now estimate that nearly four million people died of starvation between 1931 and 1935 as a direct result of state policies).

6.1 The case for multiple, not infinite, truths

The optimists believed that the commitment to truth that is central to science would spread into politics and government. Instead, the twenty-first century showed that truth is more than ever a battleground. On one side is the idea that truth is an option; that we each have our own truths; that there are no absolutes and no invalid perspectives. This idea has permeated the discourses of power, sometimes justified by a democratic spirit of breaking down hierarchies and dethroning the experts, and sometimes justified by culture wars. It stands as the anti-thesis of the scientific approach.

In the US context, this new ethos was summed up in the famous comment attributed to Karl Rove, President Bush's adviser, during the first Gulf War, who challenged people who 'believe that solutions emerge from . . . judicious study of discernible reality' and offered his alternative view: 'when we act', he said, 'we create our own reality . . . we're history's actors . . . and you, all of you, will be left to just study what we do'.[1]

Twenty years later, this flexibility with the truth had hardened. During the COVID-19 pandemic, surveys in the US showed that political beliefs correlated with perspectives on science, from views of lockdowns and social distancing to willingness to take vaccines (with a roughly linear relationship between vaccine scepticism and votes for Donald Trump), as homophily in social media constructed largely disconnected truth worlds.

In Russia, Alexander Dugin, Putin's ideologist and a hater of 'liberal modernity' and rights-based individualism, situated similar views in the context of European philosophy, claiming that 'writers like [Foucault,

Deleuze, Lacan, Derrida] recognize modernity as a perversion which is based on nothing . . .' endorsing '[Bruno] Latour and . . . his criticism of modern science . . . Russia is the keeper of the alternative order, which is its katechontic destiny'. Katechon, which is a Greek word, means something that doesn't allow reality to fall into the abyss. It is 'that which withholds'. It was associated with the Roman Empire and with the Church, the powers that resisted the tendencies to chaos, which today might include scientific reason. Vladimir Surkov, the Kremlin's in-house ideologist (with whom I once spent some time)[2] tweeted that 'it is precisely in these moments that the contradictions of ever-churning narratives, like celestial bodies in collision, produce such massive friction as to allow truth to form itself in the plasmic heat of conflict . . . really I am a pyromancer, a keeper of this flame from which new realities are born. Is there anything truer?' In this view science has no independent authority, no unique claim to truth. Sovereignty and the interests of the state precede and supersede science and bludgeon its truths with their own.

In itself this is nothing new. The cynical stance was summarized by Napoleon's sometime foreign minister, Talleyrand: 'the truth is whatever is plausibly asserted and confidently maintained'. Many contemporary leaders have followed his example, repeating obvious lies with such confidence and bravado that many believe them. We are used to 'policy-based evidence'; selective use of information of all kinds; data that is tortured to confirm any belief; the hope that faith alone is enough to achieve any goal – from the Maoist slogan 'however much we can dream, the land will yield' to American confidence that visualizing an ambition will allow it to be realized (in Oprah Winfrey's words, 'Create the highest, grandest vision possible for your life, because you become what you believe.')

The political philosopher Leo Strauss concluded that lies are essential to politics – from religion to national myths that justify one nation's arbitrary hold over land, and he took a dim view of science.[3] Goebbels went further, arguing why it was 'vitally important for the State to use all its powers to repress dissent; for the truth is the mortal enemy of the lie – and thus by extension – the truth is the greatest enemy of the State'.

Many other polities subordinate truth to meaning. Belonging trumps verification. Identity trumps doubt. And, while there are some signs of correlations between trust in government and trust in science, and that governments that are more science-like (the ones that are explicitly committed to truth, admitting when wrong, humble in the face of uncomfortable facts . . .) are trusted more, there are too many counter-examples of states that use control of the media to get away with lies and still be trusted.

Some believe that hiding truths, and not just inventing lies, is essential to the malign operation of power, and that good power is by definition open: Rosa Luxemburg commented that, 'if everyone were to know, the capitalist regime would not last twenty-four hours'. In this spirit many have tried to pull back the veils of power. Investigative journalism exposes hidden truths, connections, secret bank accounts. But, seen from the twenty-first-century vantage point, the worrying conclusion is that everyone can know, but not care. Knowledge alone is not empowering, and consequently secrecy may not be so essential for dictators as it appears.

One possible future brings fragmenting, decaying and decomposing truths, with science losing any unique authority. Any attempts at correction or fact checking simply reinforce resistance. Technologies that enable fakes and deep fakes of all kinds guarantee an arms race between claims and verification in which the verifiers will always be too late. Meanwhile, the power of 'confirmation bias', our ability to find information that confirms our views rather than challenging them, turns out to be even stronger among the highly educated than less educated.

But this picture is not the only plausible one. Wiser governments, however ideological, have always realized that their own truth is of little use if it doesn't work: its autonomy is limited. The Nazis, for example, contemplated doing without the 'Jewish science' of quantum physics and relativity but accepted their best physicists' arguments against such a self-defeating dogmatism.[4] There is a revealing story from the USSR at the height of Stalin's era that confirms the point. Lavrenty Beria was the official curator of the USSR atomic bomb project whose scientific director was Igor Kurchatov. 'Is it true', Beria asked, 'that we have to refuse quantum mechanics and the theory of relativity because they are idealistic?' (at the time, 'idealism' was the greatest ideological sin, the opposite to dialectical materialism). Kurchatov answered: 'if we reject them, we reject the bomb', to which Beria responded, the bomb is the most important, everything else is rubbish.[5]

The relationship of organized religion to science is also more nuanced than is sometimes presented. The Catholic Church persecuted Galileo and prohibited many texts (including Aristotle's). But it also supported figures like Roger Bacon, Nicolas Oresme and Roger Grosseteste (both a bishop and a great pioneer of experimentation), and the Jesuits were pioneers in astronomy (35 craters on the moon are named after Jesuits).

Power needs truth and cannot afford to be too casual. Yet recent history suggests that the status of truth depends both on the pragmatism of power and on the strength of the institutions dedicated to it. If powerful institutions are created to defend it – from public service broadcasters and fact-checking web services to universities and commissions – truth can flourish. If powerful institutions neglect it, it withers. This is the lesson of at least one strand of recent history. In the latter years of the twentieth century truth commissions proliferated, first in Latin America in countries such as Bolivia, Guatemala and Chile to uncover human rights abuses under dictatorships, later, and perhaps most famously in South Africa after the fall of Apartheid and then in countries as varied as Canada, Germany, Ireland and Norway. These official exercises to establish a shared truth were all responses to perceived wrongs, particularly against indigenous peoples, and all believed that the search for truth was an essential feature of a decent, ethical society.

As I show later there are now many examples of institutions that aim to serve both science and the state, institutions that re-assert truth as, if not absolute, then at least describable and searchable. The Netherlands has over one hundred bodies advising on science;[6] the UK has an Office for Budget Responsibility, publicly commenting on the government's spending plans, a Government Office of Science, numerous advisory committees and NICE in healthcare, which publicly rules on the effectiveness and cost-effectiveness of treatments. A majority of OECD countries now have long-term fiscal sustainability statements and independent reviews of their forecasts. In the US, there is the Evidence-Based Policy Act; in the European Union an array of committees, bodies and advisory systems (I look at these in more detail in Chapter 9). These provide advice, steers and pointers. They don't usually offer singular truths but rather recognize ranges – seeing truth not as a single thing but rather as a field of probabilities and possibilities.

6.2 *State and science and the dialectic of master and servant*

Harry Collins and Trevor Pinch described science as like Rabbi Loew's golem – powerful, willing to follow orders, but clumsy, dangerous and able to destroy its masters. This pattern is well described in the story of Frankenstein's monster and innumerable others. Here, I suggest a similar framework for making sense of the position of science, which also acknowledges the ways in which science can usurp its masters.

In his book *The Phenomenology of Spirit*, Hegel tells a complex story of the relationship between a master and a servant. There are many dimensions to this story but at its heart is a shifting relationship of power and dependence. At first, the servant works for the master, who has all the power and the money: the relationship is asymmetric and unambiguous. But, as the servant does more and more work on the world, engaging creatively with it, he becomes more aware, more self-conscious and more powerful. Meanwhile, the master becomes ever more dependent on the servant, not just for food and practical support, but even perhaps for company. As a result, the power dynamic between them transforms and turns on its head. Hence the description of this as a dialectic. It is a continuous process of redefinition as opposites reshape each other. It's not that the servant becomes a master, but rather that the servant takes on some of the characteristics of a master, and the master takes on some of the characteristics of a servant.

This framework has parallels for the world of science that has become massively larger and more powerful mainly thanks to the patronage of states which saw it as a servant, there to serve the state's need for armaments or infrastructures, or to grow new industries. But this experience of working on the world greatly strengthened the muscles of science: its institutions, skills and confidence burgeoned. It learned that it could reshape the world – literally mastering it.

States advanced too with new methods for policy, administration and regulation. But they didn't advance at the same pace as science, in part because they didn't apply comparable methods. They lacked an equivalent commons of knowledge; they rarely used experimental methods, or methods such as peer review. They could benefit from some cumulative learning about how to act on the world, how to plan, devise and implement, but there was not even a hint of the systematic work done in the main sciences.[7]

And so an ever greater imbalance has become apparent whenever a scientist speaks to a politician or a bureaucrat. One sits on top of a vast empire of thought and learning. The other is an improviser, much less able to say: if x then y. The result is that science now has an authority, even at times a sovereignty, distinct from and sometimes in competition with the classic sovereignty of the state. States depend on science – to solve problems, generate wealth, guarantee security, and to give the state legitimacy. Yet the state increasingly struggles to understand the creature it has nurtured, which has grown large, intricate and self-confident, while the confidence of politics has waned.

The opacity and complexity of science has widened this gap. With the arrival of quantum physics a century ago science became intrinsically hard for even the smartest everyday intelligence to grasp. Very few can explain how a language translation algorithm works; or the mechanisms of epigenetics. These are hard even for the scientists to explain. No wonder Arthur C. Clarke commented that any sufficiently advanced technology is indistinguishable from magic.

So the gulf has widened, and will continue widening, with politics running to catch up with the servant it nurtured, fed and funded. It has been a commonplace of political theory, at least in the West, that politics should be supreme over all else. This was the argument of Plato, restated in different forms by Lenin, Hannah Arendt and so many others. In their view politics has to be the guardian of the interests of the polis, the community, and every other field of life must be willing to concede to it. But if this superior force has inferior knowledge then its legitimacy inevitably frays. If that theoretically superior force lacks any cumulation of knowledge, if it doesn't professionalize, systematize and improve its capacity to think; if, indeed, it is locked into evidently inferior systems that privilege crude messages over sophistication, gut over head, identity over intelligence, then their relationship is bound to be unstable.

At the dawn of the modern era Montesquieu famously commented that 'constant experience shows us that every man invested with power is apt to abuse it, and to carry his authority as far as it will go. [. . .] To prevent this abuse, it is necessary from the very nature of things that power should be a check to power.'[8] Now that science has acquired power, this same point matters for its governance too. Scientists need checks, just like any other group with power that is at risk of being abused.

The problem has been apparent for more than half a century. In 1961 the soldier-turned-President Dwight Eisenhower commented in his famous valedictory speech that 'research has become central . . . more formalized, complex and costly . . .' and warned that 'public policy could itself become the captive of a scientific-technological elite' as the 'solitary inventor tinkering in his shop has been overshadowed by task forces of scientists in laboratories and testing fields'. Yet no major state felt it could risk taking another route. All continued to spend more, to build up their scientific-technological elites, to make themselves more dependent. And few attended seriously enough to the task of helping the world of politics and government catch up.

This makes it hard to think of the relationship between politics and science as one of control: of a master, whether standing in for the public or for the interests of a class or state, driving and directing a pliant servant. A good illustration is the remarkable rise of large language models, from BERT to the successive GPTs, with their vast implications for everything from education to the arts and science itself.[9] Few doubt their significance, and the risks they bring along with the opportunities. Yet, even when the scientists closely involved called for rules and regulations, governments had little idea how to respond (China for example proposed banning LLMs with 'any content that subverts state power, advocates the overthrow of the socialist system, incites splitting the country or undermines national unity', but this is precisely the kind of rule that is hard to implement in relation to an LLM).

These extraordinary achievements of science – and they are truly extraordinary – can easily dazzle. I have seen many politicians, particularly ones with little background in science, become bewitched by what to them appears a species of magic, and one that they want a part of. But patchily informed bedazzlement is not a good basis for making decisions on how to use these tools, how to encourage accuracy and truth, how to determine their uses in schools or colleges. On all of these, the supposed masters are rendered mute, reduced to open-mouthed admiration and private anxiety.

7

Clashing logics

> There is nothing more confining than the prison we don't know we are in.
>
> Shakespeare

I first became interested in the clashing logics of science and politics when I started running teams in governments. Our job was to come up with good strategies which were founded on evidence and practical to implement. The topics included drug addiction and crime; energy policy and climate change; obesity and fitness; skills and fiscal reform and we were helped by teams that typically mixed civil servants with secondees from universities, business and civil society.

Often a scientist would be seconded in. They were usually bright and diligent. But when asked to do a rapid review of evidence (much of the work in government had to be done fast) they struggled. As they researched, they became ever more aware of what they didn't know. And so invariably they asked for more time, so that they could be more thorough and comprehensive.

But it usually became clear that there was no end in sight. Each acquisition of new knowledge simply revealed other related knowledge that needed to be understood before conclusions could be reached, particularly when we were dealing with complex, multi-dimensional issues. Our scientists struggled even more with prioritization, again for the purely rational reason that they didn't have enough knowledge to definitively advocate for one option over another. The civil servants and political people by contrast loved this kind of choice and were perhaps unduly relaxed about the limited knowledge they had to help them make these choices.

This contrast between the ways of thinking stayed with me. I had been brought up to believe that a more evidence-based, science-based approach to government had to be a good thing. But these experiences suggested inherent tensions in the idea of a more scientific approach to government.

Many politicians assume that more use of science will make their jobs easier and that it will protect them from confusions and chaos. If they just follow the evidence or the science, the choices and actions will become more straightforward.

Yet my experience is that this is rarely the case. Often the opposite is true. The more deeply you engage in a topic the more you appreciate the tangents, the side angles, the gaps. Take, for example, a promising new way of teaching mathematics in school. It may do well in randomized control trials. But that could be the beginning not the end of the process of discovery. It may work for averages but not for outliers – and its effects will be disguised if there isn't the right segmentation. It may heighten stress. It may be good for exams but less good for understanding. This journey of expanding awareness of ignorance is entirely healthy and is the corollary of the journey of expanding knowledge. But it can be unsettling.

7.1 Knowledge, logics and cultures

Complex societies rest on a division of labour, a multitude of distinct roles. They have been much analysed by economists ever since Adam Smith described the division of labour in a pin factory, which made its efficiency possible (though it's an irony of intellectual history that he had never actually visited such a factory – but drew on others' descriptions). By specializing in distinct, complementary tasks, we are able to do much more complex things, and one way of measuring the state of development of an economy is the diversity of tasks it can do.[1]

But societies don't just depend on a division of labour. They also rest on a division of cultures – distinct but complementary ways of seeing the world. These can be found in roles from builders to auditors, police to teachers, entrepreneurs to engineers. These each have their own view of what is – how the world works – and their own view of what matters. They tend to have very different views of risk; different views on the balance of formal and tacit knowledge (or to put it another way, a different view of how much books matter). Some live in very hierarchical worlds, others in much more anarchic ones. Some defer to tradition, others hardly at all.

Sometimes their culture fits well with the social view of their task – we want doctors who believe in curing their patients and who live out their

Hippocratic oath not to do harm. We want cautious auditors and bold entrepreneurs, tough brave soldiers and sensitive, creative artists.

But sometimes their culture clashes with the view of society, as when accountants become too creative, or the military start running factories and acting as entrepreneurs, or teachers prioritize their own convenience over that of their pupils or doctors do unnecessary operations to maximize their income. Indeed, the cultures of distinct groups are always some combination of service to the greater whole and self-serving, with, at the extreme end, corruption and predation.

These complementary cultures have been much analysed in anthropology (for example in Mary Douglas' grid/group theories), in political science (for example in Michael Walzer's work on pluralism), in social science (with Niklas Luhmann's theories on sub-systems) and in philosophy (as in Alastair MacIntyre's work on how morality is embedded in practice). All have shown that complex communities don't work well if everyone has the same moral outlook. Instead, their 'moral syndromes' differ and need to differ, for their tasks to be performed well.

I find it useful to characterize these different approaches with the language of logics (though there are other words that capture much of the same idea, from frames and schemas to ethos). These logics guide how decisions are made and how they are justified. They are 'shared, internalized, cognitive structures which are fundamentally evaluative in nature':[2] ways of judging, differentiating and deciding.

Such logics were theorized by a sociologist, C. Wright Mills, who defined them as an 'internalized organization of collective attitudes'. They are used by people to justify themselves 'to the members of [their] universe of discourse' or to ourselves – the 'generalized other' by whom we are 'restrict[ed] and govern[ed]', since we constantly 'converse with ourselves in thought'.[3] They are inherently social in nature, only making sense within a community of practice. Another sociologist, Pierre Bourdieu, wrote of them as the 'organization of all thoughts, perceptions, and actions by means of a few generative principles, which are closely interrelated and constitute a practically integrated whole'.[4] He thought that every field implied its own logics. [5]

The world is made up of such thought logics – patterns of thought often linked to action that have their own structure, system and views of causation. They can be understood as living things – since they evolve, seek resources, spread and sometimes decay and die. Indeed, this is a helpful way to think of everything from organized religions to craft skills, parenting methods to political ideologies. They make sense in their own terms and often have their own validation and external feedback if people see them as useful and they attract adherents. And precisely because they are living things, their definitions and meanings evolve.

For my purposes three broad logics are relevant to the interaction between science and politics: those of the scientist, the politician and the bureaucrat. These are, roughly, the worlds of knowledge, power and order. They are often in tension with each other; often prone to mutual misunderstanding; but also often able to cooperate. At the end of the chapter I describe the emergent logics that may be particularly useful for future syntheses of science and power.

7.2 The logic of science

Let's start with the thought logic of science itself, which Georg Simmel summarized nicely. First people knew in order to live. Then people lived in order to know.[6] These are the logics used by the world's eight million or so scientists (it's less than two centuries since the word was first used) in laboratories and universities, with instruments, computational power, colliders, stored knowledge and living knowledge.

There is a substantial literature on what are taken to be the philosophical assumptions of science, the heart of its logic: a world that is coherent, comprehensible and obeys immutable laws; a world that we can try to sit outside, and therefore view objectively; a world that is logical and coherent. In the words of the philosopher Thomas Nagel, science aspires to the 'view from nowhere', and what it sees it expects to make sense. Natural science seeks to trace obscure facts back to clear ones (by contrast, metaphysics often seeks to trace clear facts back to obscure ones).[7]

Because of these assumptions science can be organized around a hierarchy that stretches down from the most abstract to the most practical, with the highest levels of science working on abstract theories from which other knowledge can then be deduced. Physics is the purest example of this, hypothesizing patterns or particles and then later discovering them.

The view of a logical, coherent natural world justified the distinction between basic research, the discovery of fundamental principles, and applied research which would apply those principles. Most scientists believe in what has sometimes been called 'naturalism': the idea that there is a reality, a nature, that is subject to universal laws, and that the task of science is to represent that nature, albeit imperfectly. This task differentiates science from religion, myth, wishful thinking and illusion (even if these sometimes overlap, as in the case of Georges Lemaître, the priest who was one of the originators of the idea of an expanding universe and the 'big bang').

But science is not a description of reality so much as a description of a description, everything seen through layers, layers of theory, of ways of seeing, choices about what to notice and what not to notice, choices of where to put boundaries and classifications, with misrecognition and partial understanding the norm not the exception. It is our nature to see through models never directly, but rather through sensors, mathematics, experimental exploration, hypothetical construction of analogical models, use of categories and typologies, and statistical methods to spot probabilities and regularities. Science is limited by what it can observe and a long tradition of philosophy asserts that the best it can do is fit observations to theories, rather than describing a reality that sits behind the observations. The representation or map is very different from the thing being represented. Their truth is not absolute and not all claims can be falsifiable (much of science seeks out new data to confirm theories rather than accepting disproving data as the end of the matter). Instead, scientists generally seek the truest options – not absolutely true ones, which is why a single disproof is not enough to undermine a scientific belief, contrary to what was claimed by Karl Popper. Rather inference guides to the best explanation, which is sometimes the simplest explanation, but not always (parsimony is a healthy disposition but not a healthy rule).

Siddhartha Mukherjee writes well of the way in which scientists live, rather like novelists, surrounded by ghost-like characters, some of which 'like Type I Interferon, the toll-like receptor, or the neutrophil – are mostly visible, except they exist in the half light of visibility. We think we know and understand them, but we don't, really. Some only cast shadows. Some are completely invisible. Some mislead us about their identities. And there are others around us whose presence we cannot even sense.'[8]

But, whatever methods are used, a shared hierarchy of values guides the logic: in the world of science what matters is knowledge. Its pursuit is an end in itself, whether the means is abstraction or engagement with the working world. It does not need external justification, even though it may be useful to show how, for example, a particle accelerator may contribute in the future to economic growth or space travel.

This is why the physicist Richard Feynman said that the philosophy of science is as useful to scientists as ornithology is to birds. In his world view science should be self-sufficient. It should strip accretions from other domains – the representations of myth, magic, gods or the claims of philosophers. The scientist's job, instead, is to lay bare the very nature of nature, to understand it as it is without ornament.

Yet this approach is never quite feasible, and if birds could understand ornithology it might be useful to them in avoiding predators. Moreover, the philosophy of science can help scientists see patterns that might otherwise be invisible to them. Feynman's disregard for values – and for arguing about what matters as well as what is – now looks like a symptom of a problem rather than an answer to it, but his was at least a clear exposition of science's core logic.

This, the self-referential value system of science, implies that the most admired science is science that produces the most new science. These are the rare moments when science brings a leap in paradigms – from Copernicus to Darwin to Einstein – or the invention of a new sub-discipline, generating and expanding what Husserl called the 'horizon of possibilities'. As one recent survey put it, 'for research the ultimate meta-goal is the surprise that changes the way we think and do and that leads to paradigm creation and destruction'.[9]

In short, the logic of science asserts that we live in a world that is amenable to understanding and action, that is logical, describable and representable. And it asserts that new knowledge is good in itself and that this justifies the openness, discovery and criticism that contributes to new knowledge. Other goals may be important – but they are secondary.

These beliefs help the scientific community to police its boundaries – as when a scientist is shown to have faked results or worked without sufficient methodological rigour. Within the logic of science these policings are more important than other ones – like stopping science that could create harms or that is likely to have no useful impact.

The logics of science have sprawled far beyond the boundaries of science itself because of their usefulness. A good example is the spread of what's called A/B testing in business, where companies test out alternatives with large groups of customers. Another example is evidence-based public policy, itself influenced by the rise of evidence-based medicine in the 1970s. Or there's the influence of science on personal lives as ever more people try to shape their diets or fitness with data and feedback, devouring the latest research. But at the core is a value system in which new knowledge is pre-eminent, and restrictions and restraints are suspect.

7.3 The logic of politics

Our second logic is that of politics, which I understand in Aristotle's sense of being about how we achieve the good life for our community.[10] The political view of 'what is' can be summarized quite simply. The world is made up of issues that matter to the community: these are either problems or opportunities. In response politics generates answers – policies, actions and remedies. This is its currency, a currency of diagnoses and prescriptions that are selected from the infinite range of possible priorities for attention, and that then involves both nominating issues and the people to handle them.

The nomination of issues to care about – a topic discussed by John Dewey in *The Public and its Problems* – often involves science: campaigning to take AIDs seriously in the 1980s; campaigning against GMOs; campaigning against nuclear power; or campaigning to give patients access to new cancer drugs. All are examples of the political process of saying that something matters and deserves collective attention.

But politics is also about people – who should rule – and the nomination of people has little role for science. Instead, the politics of personality, concerned with rises and falls, motives and characters, is essentially pre-scientific and so, where science sees a world that is explainable, cool and detached, politics can seem hotter, more mercurial, more fluid. This combination of people and issues then guides the narratives of politics, dealing with a public with limited willingness or capacity to follow details, and emphasizing the primary colours of ideas and identification, rhetoric, declarations, arguments to destroy a competitor. Facts and evidence are primarily ammunition for rhetoric, rather than the other way around, and politics is very much about talk: arguing, explaining, denouncing and outdoing competitors.

Indeed, because talk is cheap, 'the politician has to be able to say more than he can do',[11] and so politics often puts a burden on the state and bureaucracy that is beyond its capability, setting in motion cycles of disillusion and disappointment. The amplification of talk arises because politics is organized around competition – and in this way mirrors science. The claim is that good results come as a result of ferocious competition: 'the democratic method produces legislation and administration as by-products of the struggle for political office';[12] just as economic progress is a by-product of the struggle for profit in capitalist market economies and scientific progress is a by-product of the struggle for prestige between scientists.

Since that landscape of attention, issues and people shifts, constantly, politics has to be flexible and this flexibility is at the core of its moral position, even though politicians can be dogmatic and rigid. It is also flexible because it concerns, as Walter Benjamin put it, 'the art of thinking in other people's heads'. Indeed, the flexibility of politics – able to adjust to any circumstance – is its greatest strength (and as we shall see, sometimes a great weakness too). The longest-lived parties – the British Conservatives, the Liberal Democratic Party in Japan, and the Chinese Communist Party – survive because they can shape-shift in any direction. There may be some apparent dogmas, and these may even be incorporated into rituals. But they are all disposable. Science by contrast is dogmatic, at least in relation to methods, organized around a credo, clear on heresy and its boundaries.

François Mitterand famously said that the essential quality of a politician was indifference, echoing the Jesuit idea that we should act *perinde ac cadaver*, in the manner of a corpse, and make a difference by remaining indifferent.[13] Max Weber, in his talks on politics as a vocation, put a similar idea more positively, speaking of three qualities that are 'decisive for the politician: passion, a feeling of responsibility, and a sense of proportion . . . the decisive psychological quality of the politician: his ability to let realities work upon him with inner concentration and calmness. Hence his distance to things and men.'[14]

This flexibility – the pull of the moment – brings in another difference. Politics operates in the here and now, talking to people who are aware of their mortality, the finitude of their lives and who need answers in the present. Science solves everyday problems but looks to the infinite, to long-time horizons, offering gifts to a shared body of knowledge. As Francis Bacon wrote, *progressus scientiarium* is unlimited by time, unlike political knowledge which is contextual. As a result, science offers little consolation – little of what religion gave to the frightened, or the consolation that politics offers, that even if your life is unhappy and stunted you are part of something greater, like a nation with a destiny that makes the tribulations just about bearable.

The German political thinker Carl Schmitt tried to find the equivalent in politics of the distinction between good and evil in morality, ugly and beautiful in aesthetics, profitable and unprofitable in economics. He concluded that 'the specific distinction of the political, to which political acts and motives can be traced, is the discrimination between friend and enemy',[15] and this conclusion makes some sense. Politics is certainly animated by enemies.

But this is only part of the story. Politics is also about shared intelligence in a place and time: it is about 'being-in-the-world' to use Heidegger's formulation, concerned with home in its widest sense, a here and now, an 'us', that may or may not need an enemy to be coherent. It contrasts with the many philosophies that encourage detachment, wandering and weightlessness.[16]

So politics, unlike science, can be attuned to daily life. Politics is concerned with structures of feeling and answers to anger, and the humiliation that often lies behind anger, giving it expression, making sense of it and directing it. That sometimes makes it anti-science, and often makes it anti-technology.[17] But it can just as easily glorify science (generating the vast wealth needed for glorifying projects like the moon landings) or take credit for scientific successes (like vaccines).

This very flexibility can be unsettling. An oddity of politics is that its discourse is sometimes distrustful of itself. When a politician makes a political point about a disaster (a nuclear power station meltdown or a train crash for example), arguing that these signal carelessness and a lack of investment by an opponent in government, they are accused of playing politics. Politics is organized hypocrisy and as a result sometimes despises itself.[18] The President of the European Commission, Jean-Claude Juncker, shared a commonly held view among his peers, another version of this self-loathing: 'We all know what to do, we just don't know how to get re-elected after we've done it.'

Politics depends on the ability to swallow toads without making a face as Carlos Fuentes put it – to deny the obvious in order to sustain a narrative or argument. When huge queues appeared at Dover, the UK's busiest passenger port, in the wake of Brexit, the leading politicians had to say that this had nothing to do with the project they had advocated and overseen. When forest fires swept large areas of Australia or the US, leading politicians had to assert that these had nothing to do with climate change. The narrative supersedes everything, however inelegantly.

This takes us to the verification principles of political logic. There is no easy way to verify a political claim. We cannot prove it true or false. Instead, we use proxies. One is the quality of the individual – their trustworthiness, which may in turn depend on another proxy: are they like us, do they seem authentic or plausible? Another proxy is the logic of their argument: does it hold up, at least well enough for us? A third is experience: have they in some other role showed that they can perform well enough as a mayor, a minister or a governor?

These are all weak methods of verification and they are time-limited. A politician or a political claim may be valid now: but what about in ten years or when conditions have changed? This tendency within politics has two consequences. One is a cyclical pattern. If rulers, parties, or governments tend to atrophy, stagnate or to become even more self-serving, then it follows that they need periodic culling or refreshing.[19] The second consequence is that within politics what is often admired most is, mirroring science, generativity. But this time, the generativity comes within the logic of politics itself: the leaders who can survive and thrive, reinventing themselves, producing new polities and constitutions are the most admired.

7.4 The logic of bureaucracy

A third logic is to be found in the everyday rhythms of government and bureaucracy, a layer below the leaders. Here are the administrators, managers, civil servants, bureaucrats and functionaries. In ancient Sumeria they were the managers of grain. In the British empire the district officers; in contemporary China both the local party secretary and the local mayor. They have to carry out what politics demands and to cope with the likely gap between what it demands and what the state can deliver.[20]

They generally seek order and stability. They are judged by outcomes – safety, welfare, the absence of crises as well as by everyday progress. Their lives are shaped by rules; written instructions; reports and records. Law sets the parameters of their work. They work with representations of the world – maps, censuses, data – and then seek to influence it, either towards a new equilibrium (such as lower crime) or towards a direction (growth of population), sometimes using explicit orders, sometimes deliberately ambiguous ones.[21]

The moral world of bureaucracy is one of ordering what otherwise drifts into chaos. Bureaucrats see a world without them as naturally disorganized, conflictual, inefficient and unhappy, a world of capricious acts and abuses. The first time I spent time with Chinese government officials I remember being astonished at how starkly they thought the public they served were just a step away from unruly anarchy, quite unlike the Western caricature of a population of dutifully compliant civilians, moulded by millennia of Confucian deference.

The world such bureaucrats try to govern needs rules and predictability. In principle, too, it is also a world made up of roles that are distinct from the individuals who, temporarily, fill them. Of course, some bureaucracies become opposite to this: they can be corrupt, arbitrary and capricious. But it is striking how many autocracies and kleptocracies feel the need to maintain at least the illusion of the rule of law and rules, of power that is impersonal.

In Hannah Arendt's words, 'in governments by bureaucracy decrees appear in their naked purity as though they were no longer issued by powerful men but were the incarnation of power itself and the administrator only its accidental agent. There are no general principles which simple reason can understand behind the decree, but ever-changing circumstances which only an expert can know in detail. People ruled by decree never know what rules them because of the impossibility of understanding decrees in themselves.'[22]

But the deepest deep state is oddly not the one concerned with security and conspiracies but rather the state of the mundane and the prosaic. This is where bureaucracy comes into its own, and delivers sharp falls in child mortality or rises in GDP. Technocracy is one extreme of the bureaucratic logic applied to a society. It can be efficient, indeed surprisingly effective. Technocracy gave many countries the most successful periods of their history: France, South Korea, Japan and China, achieved economic growth and rapid improvements in standards of life during periods of technocratic rule. But technocracy is also prone to vices: seeing people as means not ends, lacking empathy and so missing emotions that matter. It can all too easily be manipulative and too tied into the interests of other elite groups.

Bureaucracy often feels that science is a natural ally: equally cool, rational and rigorous. But the interactions often bring friction, since science is so hard to plan or steer. Indeed, the biggest dilemma for bureaucracy is how to scale and make systematic something so fluid and unpredictable as knowledge; how to fatten the golden calf but not kill it with too much attention. Thomas Hughes commented decades ago that 'a mission-oriented laboratory tied to an industrial corporation or government agency with a vested interest in a growing system nurtures it with conservative improvements . . .'[23] and there are many examples of well-intentioned bureaucracy stifling creativity. Lao Tsu's argument that governing a great country is like cooking a small fish – 'don't overdo it' – can also serve as an injunction to anyone seeking to make a policy for science.

7.5 How the logics intersect and clash

So how do these different logics interact, intersect and conflict? The short answer is: with some difficulty and mutual misunderstanding. Niklas Luhmann was surely right to argue that all sub-systems represent the world in their own terms and translate the logics of other fields into their own logics. Economists see a world made up of prices and incentives; lawyers a world of compliance and crime; the media a world made up of stories, and so on. These self-referential logics give us 'art for art's sake' and 'business is business'. Politics translates science into missions, myths, stories and glories. Bureaucracy translates science into laws, rules, programmes and inspections. Conversely, science pictures an idealized world of government entirely based on evidence, experiments and facts.[24]

The three logics sometimes intersect. The most important field of intersection is risk. Each of the logics has a language for risk and is at ease discussing it. Politics exists to protect people and so needs science to advise on risks – predicting floods or earthquakes. Bureaucracy establishes rules, protections for health and safety, inspections and safeguards. There is a similar overlap around opportunities – a new way to create clean water or cure a disease is a natural space for politics, bureaucracy and science to share thoughts and options. War and economic growth are also comfortable common grounds.[25]

The logics can also overlap around shared narratives. Political stories give an umbrella to science, a way of motivating and guiding. These have included narratives of progress, the Cold War and Space Race, social problem-solving and mass health, zero carbon and more recently the Sustainable Development Goals (SDGs). This is where missions come in – and explains why they are so appealing to politicians. They turn the messy, complex and unpredictable world of science into a much simpler political narrative (though implementing them is infinitely harder, and far more missions are launched than ever concluded).[26]

There are also many fields where the logics diverge. I've already mentioned a few: how scientists struggle with trade-offs which are normal in politics and economics, like putting a price on a human life. Another divergence is breadth. Scientists usually limit themselves to what they know and understand, partly out of deference to other scientists. A few late in life spread their wings and are often shot down by their colleagues. But most are instilled with a doctrine of limitation: speak only about what you know.

Politics is in some respects the opposite. It is a Swiss army knife, the universal, cross-cutting power. It can pass laws on anything, from the price of butter to actions in the family home. Politicians are required to have opinions on everything, including things they cannot possibly understand let alone influence, and it is a brave one who admits too often to ignorance or impotence. Yet they are rarely capable of living up to this role and so live in a state of anxiety, a perpetual imposter syndrome.

Even more fundamental than this is the divergence around flexibility. The price politics pays for its flexibility is, paradoxically, that it doesn't advance. It has little cumulative knowledge to draw on. Each generation invents new political rules for itself. Few politicians read political science, and even those who do, don't use it to guide their decisions. By contrast, the price science pays for science's solidity and dogma is a lack of flexibility, and sometimes an inability to communicate or empathize.

Yet another tension is the role of facts or truths. Science may never find final truths but it believes itself to be on a journey towards them. In politics there are no settled truths of any kind. Everything is contingent and contextual. One popular view says the post-modern turn means any founding of democracy on fact or science is impossible. Perception, subjectivity, the interior fused with exterior: these cut against any solid truths. But while this is often true of politics it is not the whole truth. Reality impinges despite the great capacity of individuals and communities to hold on to myths and fantasies. Flying a plane, driving a car, building a bridge or ending a pandemic – all involve material realities. So do wars. Reality bites and sets limits.

7.6 Impure philosophy

What of philosophy – can this provide a shared grounding? Philosophical reasoning certainly helps in showing up the limitations of different logics. Wilfrid Sellars, for example, wrote about how the scientific world view shows that 'the common-sense world of physical objects in Space and Time is unreal – that is, there are no such things'. Science attempts to be 'the measure of all things, of what is that it is and of what is not that it is not', yet it's bound to fail: so philosophical reflection at least helps with humility.[27]

Philosophers have also often attempted to provide a bridge between the logics. Eugenics was one of the most influential ever, a body of thinking that drew on science, was refracted through utilitarian and other philosophical principles, and then manifested in political programmes and laws, appealing to many of the cleverest minds of the time. Unfortunately, it tended to see people as means not ends, as numbers not individuals, and like many well-intentioned philosophical ideas, denied the people likely to be affected by decisions the agency to shape them, imposing instead an external supervening logic.

One of the current variants of this way of thinking is sometimes called 'long-termism', favoured by philosopher William MacAskill, and sometimes echoed by the business leader Elon Musk and others. They claim to take the long view of human history, from which vantage point all that matters is humanity's long-term future. It follows that our ethical purpose should be to maximize how many humans and other intelligent life forms thrive in the future. Today's needs are secondary.

Again, this body of ideas draws on scientific knowledge, particularly analysis of biological risks, climate change and other threats to humanity; it refracts it through a particular philosophical lens – a variant of utilitarianism – and then proposes actions. This way of thinking is logical and appeals to a certain kind of mind for its tough clarity.[28] Such thinking has a long history – ranging from Russian cosmism to the harsher end of neoclassical economics. It provides useful thought experiments in the lecture room at universities. In the case of Elon Musk, it feeds his stated vision to 'make humanity a multiplanetary species'.

The approach can be used to brush aside calls for improving conditions and alleviating suffering among the living here on Earth now because, the theory goes, giving a poor person a blanket isn't likely to be as useful for the future of humanity as building a rocket to Mars.

But it will be obvious that it has no place for the political logic described earlier: oddly, although it claims to be political in Aristotle's sense, concerned with the good life of the collective in its largest interpretation, it is antipolitical in its ethos and it lacks the means to reflect on itself through other logics. In this respect it also ignores the lessons from similar ways of thinking, which show not just how little we can know about the future, but also how dangerous it is to give the objects of decisions no say in their shape.

My own preference is for less precise thought experiments and more actual experiment. Instead of deduction, and the attempt to find general logics, I find it more plausible to see societies as much more complex than our mental models: it follows that we should try to help the world to evolve by combining imagination and testing, whether in relation to new drugs or new policies. As Kurt Vonnegut wrote: 'science . . . has been built upon many errors; but they are errors which it was good to fall into, for they led to the truth'.

Paradoxically, these philosophical attempts to bridge the worlds of science, politics and bureaucracy offer no standpoint from which to interrogate themselves, yet they do help to clarify the gap between a cool world of detached science and the warm world of human life, and they do remind us why science and logic alone cannot tell us what matters.

7.7 Is all science political?

But it doesn't follow, as some claim, that *all* science is political. To assert this is to misunderstand both. Some science, at some times, becomes political, particularly when it becomes technology. But the concept of politics is stretched too far if we extend it to academic debates about the designation of distant galaxies, the physics of subatomic particles or the behaviour of the cell. These become political when they develop in ways that have consequences for populations and societies; they become political when they bump into conflicting values and beliefs; they become political as interests organize around alternative options. But it's dangerous for politics to seek to intervene too far upstream, or in the realms of theory. The more it does so, the more it commits a sin of category error, applying ideas that should be rooted in the present, in political contexts, to ideas that are abstract and timeless.

It's more useful to think of science and politics as functioning as fields, cooperating but also competing for power and capital, and dependent on their contexts – what sits outside their logics as well as inside. The importance of these supportive fields becomes very apparent if you take a politician, a scientist or a bureaucrat out of their web of supports. A politician transplanted to another political culture or another field becomes not just powerless but also mute, irrelevant, nothing but air. A scientist transplanted into another discipline, or into a culture that has no traditions of science, is rendered equally powerless and useless. They can probably offer little help in growing food, finding water or resolving conflicts. Instead, their value depends on the system of which they are a part. Similarly, a bureaucrat without a bureaucracy is nothing.

This is why it is usually misleading to seek power in individuals. It's true that some individuals can dramatically change the direction of history, or of an institution. But, more often, individuals are better understood as vehicles, their options sharply circumscribed by their circumstances. Power moves through them as much as from them.

Moreover, that power is not self-contained or self-referential. Each field also depends on its success in persuading others to recognize it. Politicians depend on legitimacy that is provided by the public – and periodically whole classes of politician lose this. They depend on their strong and weak supporters – the people willing to take to the streets, to join their party, to confront opponents, as well as the much larger number willing, sometimes, to lend them a vote. Scientists depend on their authority in the eyes of the other fields that provide them with money, status and freedom: governments, businesses, media, philanthropy and, as for politicians, the wider public. Again, these can easily be lost and they require continual work to sustain them, to grow and nurture their claims. If scientists face too much political opposition – from campaigners demonstrating, petitioning, boycotting – their work becomes much harder and their licence to operate is diminished.

In each of these respects both science and politics are better understood as fields of relationships rather than as things, as metabolisms that depend on drawing in energies. All societies are in part 'a theatre where social agents act out their identity, strut their stuff, persuade us to believe what flatters them most, and discredit their rivals', and science, at least in its social meanings, is just one of these competing claims.[29] This perhaps explains why science has to work hard to persuade, justify and head off threats.[30]

7.8 Future synthetic logics

This matters for where I take the argument later. My interest is not so much in how an individual politician can be given some formal power over science, or vice versa. Rather I'm interested in how we can shape and design the systems that weave science and politics together: how we can make them cognitively rich, wise, looped and able to interrogate what's known, whether about the facts of the world; the imaginative possibilities; the emerging innovations; the proven evidence; the ideas that failed.

At their heart perhaps we need new logics that draw on but are distinct from the ones described earlier. They can perhaps best be described as logics of transition. We can see them emerging in much of the work underway around issues such as climate change, designing transition paths for transport, energy and housing.

For the people working in these fields what is all important are the outcomes achieved, the transitions to more effective and more sustainable systems. They then work backwards to the relevant sources of knowledge rather than forward from existing disciplines. They constantly loop between what is and what matters, between facts and values. They constantly seek to grasp the dynamics of whole systems. And they include a bias to action, willing to act with perhaps 70 per cent of the knowledge needed, rather than waiting for one hundred per cent.

PART IV

The Problem of Institutions

Solving the Science–Politics Paradox

8

Split sovereignty, or the role of knowledge in corroding the supremacy of politics

In this part of the book, I look at the interface between democracy and science and the science–politics paradox – the paradox that only politics can govern science, yet politics has to change to be able to do this well. I start with a deeper investigation of the concept of sovereignty, and why we should consider it multiple rather than singular, with claims to sovereignty for collective knowledge as well as the will of the collective in the present.

I then turn to the many ways in which democracy engages with science: the world of science advice; the attempts to integrate science into politics through the use of evidence, experiments and data. I show how the task of steering science has become harder – rather like steering a trolley with ever more items piled on top of it.

I look at how democracy could do more to shape the ends of science, which still reflect a pre-democratic bias towards state interest and commercial interest rather than public interest. I examine some of the crucial dilemmas – around pathways for technology, and answers to stagnating productivity – before setting out how knowledge commons could provide a shared infrastructure for a healthier engagement between politics and science. My main argument is that we can no longer think of science as an external input into politics; instead, we need to interweave and synthesize the two, in institutions, processes and the mindsets of decision-makers, growing the capabilities for metacognition that are so essential to the complex challenges of our times.

8.1 The nature of sovereignty

Earlier I raised the question of sovereignty: who or what can make a claim to rule? I showed that we have been brought up with assumptions of unlimited political sovereignty and, more recently, with assumptions of unlimited individual sovereignty, both of which are radically misleading about the nature of the world we live in. I've suggested that global science, now finding a voice on almost every question, is creating a new kind of political phenomenon, a claim to power from outside the system that derives from knowledge (rather as, in the past, claims to power came from belief and religion). The commons of intelligence and knowledge that science is part of has become partly sovereign in itself, earning an authority distinct from the people, nations or interests, though shared with them.

This shift – which is set to be reinforced as new forms of machine intelligence proliferate – has profound implications for the future of political theory and practice.[1] It challenges assumptions and constitutions.

But what is the sovereignty that science may have acquired. Is it just influence? The ability to advise leaders? Or is it something more? And how can politics be refashioned to become adequate to the task of shaping science, with sufficient insight and wisdom?

To answer these questions, we first need to ask what sovereignty is. It is in some respects an archaic concept. It reaches back to the claims of kings and emperors to justify their rule. Yet it remains at the heart of our political systems. Sovereignty concerns who or what can make a claim to power, including power over us and power that shapes the prospects of future generations.

The idea of sovereignty has had a twisting and complex history, which has ended up with the claim of politics to a supremacy over everything else. The original ideas of sovereignty came from religion. Since only God could be truly sovereign, the ruler had to abide by the laws of God and, if they failed to do so, they lost legitimacy and it would become acceptable to challenge them or ignore their commands. That sovereignty was intimately connected to truth since God was, by definition, true and omniscient.

But, in the early modern period, the idea detached itself from its origin. In the seminal work of Jean Bodin, the ruler is presented as supreme and self-sufficient, and in no need of compromise with Lords and others who threatened chaos (Bodin also wrote a book on sorcery, another evil that had to be driven out). Similar ideas were taken up in England thanks to the influence of Thomas Hobbes and others. Only a supremely powerful sovereign could protect society from discord and war and absolute sovereignty for the state was necessary for a people to be awakened to its historical mission.

As democracy spread, the locus of sovereignty shifted from the ruler to the people and back again. The people should be sovereign but vesting or lending their sovereignty to the rulers they elected. Their freedom was achieved through the medium of others – politicians and, in time, a functioning state, with power concentrated in parliament. In England, a generation after the King's head had been cut off and his son restored as Charles II, the Earl of Shaftesbury stated the new position: 'the Parliament of England is the supreme and absolute power, which gives life and motion to the English Government'. Two hundred years later A.V. Dicey, probably the UK's most influential constitutional lawyer, wrote in a similar vein: 'Parliament has the right to make or unmake any law whatever, and . . . no person or body is recognized by the law of England as having the right to override or set aside the legislation of Parliament.'

Marxism took this idea of a sovereign people to an extreme. The young Marx wrote that 'in real democracies the political State would disappear', an idea realized, briefly, in the hot politics and direct popular sovereignty of the Paris Commune and the early Soviets. This idea – which turns the public into a kind of god – takes different forms. In the West it makes the public heroes: the people's views trump all others. In China, that unlimited sovereignty is claimed by the Communist Party, which represents the people by virtue of its success in winning civil war and revolution. In all cases sovereign power is uniquely able to change the rules of the game – the 'state of exception', which allows the state in exceptional circumstances to suspend freedoms and laws.

In democracies the optimistic story sees progress as involving ever-expanding popular sovereignty, a progression from passivity to activity, from servitude to becoming masters, as the people of the world become more engaged, more empowered, and more aware.

The older ideas of an indirect dimension to sovereignty, and of alignment with the laws of God, were quietly dropped. Instead, sovereignty became singular, united, and absolute. It was this kind of sovereign state that then mobilized science as its aid, a tool for the pursuit of its missions: war, conquest, education, public health. That sovereignty might be *de jure* – the legal right to act – and it might be *de facto* – the actual ability to do so. But the implication was that a parliament or assembly should be able to pass laws on anything, from behaviour in the home to business and beyond.[2]

8.2 Plural sovereignty – knowledge, ecology and future generations

Now, perhaps, we are returning to something closer to the earlier idea of sovereignty. Once again, it is no longer singular but rather divided and contingent, only now our ideal of sovereignty depends on power being used in sync with collective knowledge and science, which have taken over the role once played by God. For a government to be legitimate it has to be able to demonstrate that it acts in accordance with collective knowledge. It if is at odds with them it loses the right to command and demand obedience.

This shift is far from complete. But it can be seen as an emergent property of modern polities. A growing number promise to base policies on evidence. The US Congress passed its 'Evidence-Based Policies Act' in 2018. Some have spawned networks of institutions to advise on the state of collective knowledge. Global bodies – from the OECD and World Bank to UNDP – advise on global best practice and what the evidence advises. The elected politicians have every right to ignore this evidence – which may be misleading. But they have less right to be ignorant of it and, if they do ignore it, they need to provide reasons. Here we see an emerging dialogue between two notions of sovereignty: one the familiar idea of sovereignty vested in the people, conferred on elected politicians, the other the idea of sovereignty vested in a shared body of knowledge that includes science.

This is happening in parallel with a shifting view of other claims to sovereignty. One is the claim of nature. New Zealand gave legal personhood to a national park (Te Urewera in 2014) and a river (Whanganui in 2017), in each case appointing people to act as guardians of that personhood. After thirty thousand Ecuadoreans brought a legal case against Chevron for drilling in the Lago Agrio oil field that poisoned the soil, Ecuador in 2008 included in its constitution statements recognizing 'the inalienable rights of ecosystems to exist and flourish' giving 'people the authority to petition on the behalf of nature, and requiring the government to remedy violations of these rights'.[3] The failed Chilean constitution of 2022 attempted to embed a whole series of values – from many rights (child protection to neurodiversity), energy and clean air to indigenous property. It also sought to establish nature as a holder of rights alongside various mechanisms for protecting it and penalizing damage.[4] We will, I'm sure, soon see digital intelligences as well as humans taking on designated roles to stand for, think for and represent parts of nature.

A related step confers some sovereignty on future generations: after all, why should only the people alive now have the right to do what they want with resources which long predate them and which should survive long after they are dead? Wales, in 2015, passed a Future Generations Act and created a new post of Future Generations Commissioner, charged with commenting on public policies from the perspective of the unborn. Germany's Constitutional Court ruled some climate-change policies unconstitutional because they shifted the burden to future generations. Many parliaments have committees of the future and there is growing interest in how to respect the interests of future generations, who will bear the consequences of today's decisions.

So, the claims of science sit alongside this partial deconstruction of sovereignty which is breaking down the absolutist claims of politics to supremacy over all else. But how in practice should science relate to politics in relation to everyday choices and dilemmas – whether to fund fusion projects, how to regulate new pesticides or products that claim to fuel health and happiness? And how should it relate to democracy? These are the questions I turn to next.

9

Democracy meets science

On 12 March 2020 the UK Prime Minister, Boris Johnson, appeared at a Downing Street emergency press conference, prompted by the rapidly deepening COVID-19 pandemic. On one side he was joined by Professor Sir Patrick Vallance, the Government Chief Scientific Adviser (GCSA). On the other side stood Professor Chris Whitty as Chief Medical Officer (CMO). The Prime Minister promised that 'at all stages, we have been guided by the science'. After that, the press conferences happened regularly, sometimes every day, attracting audiences of millions and providing the public with an explanation, a narrative and a strong underlying message: that the decisions were being 'led by the science'.[1]

Here appeared to be a high point of science's influence and prestige, with politicians deferring to the superior insight of the scientists. But the approach soon fell apart. The scientists talked of using 'herd immunity' but then backtracked when the public interpreted this as meaning that they were happy to see thousands die. Pronouncements that masks made little difference to transmission were followed up with laws requiring mask-wearing. The editor of the *Lancet*, Richard Horton, a month later called the belated introduction of lockdowns 'the biggest failure of UK science policy in a generation'.[2] When the government's most influential adviser (not a scientist) was shown to have brazenly flouted rules he had helped shape, there was a huge public outcry, and a sharp decline in public confidence, but no comment from the senior scientists. A year later a parliamentary inquiry was scathing of the mistakes made, of the groupthink, the narrow sources of advice and the harms which resulted.[3]

As I'll show later, the uncomfortable experiences in the pandemic showed up many of the flaws in the relationship between science and politics: the lack of any methods or even languages for synthesizing diverse issues and interests; the blurred accountability for decisions; and the ambiguity of the science itself.

9.1 Scientists advising politics: the role of the 'science triangle'

The scientists who appeared at the evening press conferences were described as advisers. Politics depends on figures like this – advisers with backgrounds in science – to guide it on how it should guide science. Many theories have tried to make sense of how this advice should be organized. They assume that sovereignty remains in its traditional model, a monopoly of the politicians, and so present the scientist not just as autonomous but as external: an adviser to systems and processes for which they bear only limited responsibility. Politicians still call the shots, but always with advice from the best scientists. George Russell, an Irish politician, in 1912 coined a phrase (often attributed to Churchill) that summed up the relationship: scientists should be 'on tap, but not on top'.

Over the last half century, such advisers have multiplied, sometimes sitting within government departments and sometimes having the ear of the Prime Minister or President. In the UK, for example, each department was encouraged to appoint its own Chief Scientific Adviser in the 2000s, linked by a Government Office of Science.[4] US Presidents have had Chief Scientific Advisers since Roosevelt's time (most recently Ari Prabhakar), who also run the Office of Science and Technology Policy in the White House. These advisers are now so well established that they have their own global network – INGSA – and a sophisticated view of how best to navigate the dynamics of real governments and politics.

Parliaments also have their own structures, and these, too, have their own networks, such as the European Parliamentary Assessment Network linking teams in 25 parliaments, which try to advise parliamentarians on important trends. Many parliaments set up Offices of Technology Assessment in the 1960s and 1970s, as well as advisory committees of all kinds, all seeking to mobilize expertise to feed into processes that are essentially political and bureaucratic. The US Congress had one from 1974 to 1995, and Germany still does. France has a Parliamentary Office for Evaluation of Scientific and Technological Options; the Netherlands the Rathenau Instituut. All told, there are over 20 such offices in Europe,[5] and these have survived the kind of political attacks that led to the demise of the OTA in the US, victim of Newt Gingrich's neoliberal revolution, which portrayed it as a bureaucratic and costly enemy of innovation.

The European Union is a particularly striking example of the accretion of mechanisms of advice. It has the Parliament's STOA (tied into a dozen or more committees and made up of MEPs) alongside the Commissions multiple advisory committees; agencies (such as EFSA, EDA, EEA, ESA and ECDC); the European Parliament Research Service; Science advice for policy by European Academies (SAPEA) as well as a Group of Chief Scientific Advisers (with seven playing the role that in some governments is given to a single scientist after a short-lived experiment with a single one). The Joint Research Centre has 3,000 staff, 1,400 publications per year, 42 large research facilities, and more than 100 economic and biophysical models, a unique model of policy-driven science seeking to answer questions in a calm, rational and transparent way.

Some nations also have strong traditions of independent, honest research and inputs. The Netherlands is a particularly good example, with a rich history of organizing science, and social science, advice into government through institutions such as the WRR,[6] perhaps helped by a political culture that has generally involved broad governing coalitions.

The best of these sources of advice aim to make their advice as transparent as possible; to be open and honest about disagreements and uncertainties; and to communicate outwards to the public as well as upwards to governments.

We can think of these often elaborate systems of advice as part of a 'science triangle'. There is a supply of knowledge and advice; a capacity to absorb that advice directly into governments and politics; and there is a surrounding environment (media, politics, civil society) that influences how supply is used.

The relationships are relatively simple when these three roughly align. The advisers become adept at understanding what advice may be used (Sheila Jasanoff, for example, wrote of expert scientific committees advising US federal agencies that became good at producing 'serviceable truths').[7] But the relationships become much more problematic when they are out of alignment, when politicians are either uninterested in, or sceptical of, advice, or when the media amplify marginal issues over big ones. You can lead a horse to water but you cannot make it drink. In the same way you can lead a government to knowledge but you cannot make it think if it doesn't want to.

So, in the standard view, the relationship of science to politics is understood in terms of inputs and advice. In principle the adviser's role is straightforward: to synthesize the best available science to help decision-makers, a role which requires them to be neutral and objective. Scientific experts, the more prestigious the better, feed in advice, ideally acting in as neutral and disinterested a way as possible and sticking to the scientific facts. Politicians listen to their advice and then add in other considerations – values, priorities, political concerns – before making judgements that guide actions, laws, investments, prohibitions and regulations. The scientists get on with their work.

Unfortunately, not much about this picture stands up to close scrutiny. A first reason is that the political and scientific are often hard to distinguish. The writer Ed Yong, who won a Pulitzer Prize for his coverage of the COVID-19 pandemic in the United States, commented that 'the naive desire for science to remain above politics meant that many researchers were unprepared to cope with a global crisis that was both scientific and political to its core. Science is undoubtedly political, whether scientists want it to be or not, because it is an inextricably human enterprise.'[8] Scientists will often seek to remain at arm's length; to deny any policy knowledge, or legitimacy. But this abnegation of responsibility may be increasingly hard to sustain.

Real scientists have opinions, enthusiasms and biases of all kinds. Few can be wholly disinterested. In a political environment, people ask 'what's in it for you?', whether in terms of their interests or their values, yet the scientific culture refuses to accept the premise of the question. Scientists do tend to be trusted. But they are less likely to be seen as wholly neutral than in the past. It's understood that they have their own beliefs and axes to grind. Sometimes we may be more inclined to trust people who are open about their biases, such as lobbyists or NGOs.

A second problem with the classic image of disinterested advice is that advice inevitably merges into decision-making, sometimes very visibly, as during pandemics when the people designated as advisers were effectively making far-reaching executive decisions. Sheila Jasanoff wrote of the 'fear of letting experts usurp that part of decision-making which should be truly political'[9] and so advocated keeping them at arms-length, as just advisers, but that assumes a capability on the part of the 'truly political' that may often be missing. An alternative view says that the scientists should be more honest about their power, and more accountable for it, not least because 'science plays an increasing role in defining the problems for which it is then called to give advice about once these problems are on the political agenda'.[10]

9.2 Brokers and intermediaries

There is no avoiding the need for intermediaries and brokers to help governments handle science, just as they need brokers to guide on how to influence the economy or big systems like education. But what should these intermediaries look like? The academic Roger Pielke has been one of the main promoters of the idea that they should try to be 'honest' brokers.

The honesty he wanted to promote was to be distinguished from 'stealth advocacy', when scientists weave political opinions into what are presented as neutral scientific judgements. As so many topics – from climate change to gender in sports – become polarized, the risk grows of 'political battles . . . played out in the language of science, resulting in policy gridlock and the diminishment of science as a source for policy-making'.[11]

As Pielke points out, though, different issues bring different types of challenge. In some cases, there is broad consensus on both the science and on the politics. This is what Pielke calls 'Tornado Politics': confronted with a natural disaster it's not so hard for science to provide advice. On much more contentious topics, like abortion, things become harder, and he argues that scientists can either try to narrow down options (as advocates) or widen them (the honest broker role), acknowledging both stakeholder interests and perceptions and what the science says.

But it's possible that scientists can no longer play the role of honest broker, or at least not in the way they once did. There are far more intermediaries – a babel of voices interpreting science, ranging from corporate lobbies to NGOs, academic groups to magazines and websites. Some of these work hard to build up a reputation for neutrality and independence. Others don't bother.

Another academic commentator on science advice, Bruce Smith, describes an event he attended with politicians (representatives in Congress) and scientists. One member of the audience asked: 'Who would you turn to most frequently for scientific advice: a committee staff member who is tracking a topic, someone from the Congressional Research Service or a Congress-wide staff agency, an outside think tank person, a university scientist, or whom?'

To the surprise and consternation of the assembled colleagues, mostly academics, the two men said without batting an eye that they would turn to their favourite lobbyist. For a thoroughly knowledgeable analysis of the issue, for a fair presentation of both sides, for singling out the central points in dispute that required the Congress to decide, and for a prompt and timely response, the lobbyist won hands down. Lobbyists have to give you accurate information, they said, or their reputations will be ruined. They have experience, they know what you need, and they will give you a pretty good and objective assessment.[12]

Since the political will in the end be political, it's not obvious that a politician should always prefer a purely neutral view to one that is aligned with their values. Meanwhile, for the scientists themselves there is no guarantee that they will like the uses that are made of their research:

> researchers are seldom successful in directly influencing policy decisions, even if they work hard to develop better relationships with policy staff. [. . .] In general, those who produce rigorous evidence and evidence-informed policy ideas do not control how their ideas are interpreted, modified and used . . .[13]

Brokers are indeed needed to translate and adapt that knowledge. But being a broker is different from being an adviser. It certainly involves being fluent in the three logics described earlier, but it often benefits more from breadth than depth: being an eminent scientist in a particular field may be poor preparation for such a role. Indeed, too much eminence can work against a good sense of who the knowledge is being produced for, and the organizational and institutional contexts where it will be used.[14]

Science advisers sometimes struggle to give a balanced picture of the many kinds of knowledge that may be relevant to a decision, as they are almost inevitably biased in favour of their own specialisms and their own networks of collaborators. They tend to privilege high status areas of science rather than the most relevant ones and are not always good at offering a rounded view, including contrary perspectives and disciplines. Moreover, surprisingly few have developed methods for synthesis (a topic I discuss in more detail in Chapter 13).

This is why I emphasize the role of metacognition in institutions, which mirrors the vital role that it plays in individual thought. This ability to think about thought, and to be intelligent about intelligence, should be essential for the governance of science and technology. It requires some familiarity with many different fields and ways of thinking, and an ability to know what types of thought are most appropriate for different tasks: why the place of science is bound to be very different in topics as varied as regulating building materials, laws for human cloning, the ethics of AI in mobile phones and children's mental health.

It follows that the role of intermediary and broker may be better performed by people with less detailed expertise in the subject matter but more expertise as intermediaries or curators, generalists, like the best science journalists, who can stand back and see the patterns more easily than someone too rooted in a particular discipline. It may be done best by people who have thought hard about how to synthesize multiple types of knowledge, and who recognize that the most relevant knowledge will vary, depending on the context. And it may be as much about navigating and mapping the ecosystems of knowledge and opinion as it is about distilling this into something called 'advice'.

A large part of this involves organizing conversation. Although we live in a world of prose reports and websites, insights are still much more easily absorbed through discussion or visits than just through reading. We engage best with ideas through having to talk about them, and having to reformulate them in our own words. This may be why a recent survey of science advice concluded that 'the most highly recommended science advice process combines analytic rigour with deliberative argumentation . . . the mutual exchange of arguments and reflections, to arrive at evidence-informed and value-balanced conclusions in a discussion'.[15]

9.3 Integrating science and politics through iteration and experiment

Such navigators can provide one kind of bridge between the worlds of politics and science. But there are many others.[16] The simplest examples come when scientists are appointed into executive roles. Here is at least one answer to scientist's disdain for politics. As Plato warned, 'one of the penalties for refusing to participate in politics is that you end up being governed by your inferiors'. Instead, the best brains, occasionally, dive into the murk of decisions and trade-offs. Two of President

Obama's energy secretaries were world-class scientists – Stephen Chu and Ernie Monoz. Monoz was able to progress the 2015 agreement to limit Iran's nuclear programme in part because he had studied at MIT in the 1970s with his Iranian counterpart Ali Akbar Salehi. President Macron appointed a leading mathematician, Fields Medal winner Cedric Lo, to advise on AI, though he struggled with the role. He also appointed a leading environmentalist, Nicolas Hulot, to be minister for ecology and inclusive transition in 2017 and charged him with leading on climate change (though he resigned in 2018, complaining he was being ignored). In these cases, the integration of science, policy and politics happened within their brains, as they had to manage fluency in multiple languages and logics, and to synthesize different ways of thinking.

Another form of integration embeds evidence and experiment into the everyday work of government. I have spent much of my life as a practitioner and advocate of this kind of evidence-based, or at least evidence-informed policy. At its simplest this aims to introduce some features of the scientific method into the daily work of government, influenced by earlier movements like that towards evidence-based medicine in the 1970s.[17]

Its premise is that before making policies governments should find out what is known, and what is known about what works. This should not be controversial. But it is rarely simple. It is not just a matter of 'what works' but of what works, where, when and how, and what might be transferable from one place to another. Contexts matter. Few policies are simply transferable, though many things have spread to a remarkable degree, from income taxes to parliamentary democracy, which was once thought quite incompatible with Hindu or Confucian cultures, and from search engines to primary schools. Humans are good at copying, though there is never a linear process from evidence and analysis to action, rather a more iterative process of adaptation.[18]

The simple idea that you should know before you act is becoming more widespread. Evidence syntheses; 'what works' centres; rapid evidence assessments – all are becoming part of the everyday life of public bureaucracies. A parallel track emphasizes co-production, the engagement of the users of knowledge in its production.[19]

I've been closely involved in setting up some of the 'what works' centres and now spend much of my time synthesizing evidence for decision-makers. This is not glamorous, but rather cautious and careful. Yet it reduces the risk of unnecessary mistakes and can, over time, achieve remarkable successes, such as the story of how evidence was mobilized to change literacy methods in England, using systematic evidence to overcome opinion and ideology, and achieving dramatic improvements.[20]

There is now plenty of evidence that use of evidence works. A review of federal legislation over the last thirty years related to behavioural health found that bills that explicitly referenced scientific evidence were more than three times more likely to become law than ones that didn't.[21] There were similar patterns in other policy areas, including substance abuse and human trafficking, and in state legislation.[22] This at least suggests that closer integration with scientific knowledge of all kinds is feasible.

Another point of connection is experiment: the idea that government works best when ideas are tested out first. Here the scientific method and mindset is introduced into the daily workings of the state. China has a very long tradition of such experiment and, in recent decades, experimented with new economic policies, social insurance and welfare. Chairman Mao promoted what he called 'point to surface ideas', an approach encouraging experiment: 'model experiences' he wrote 'are much closer to reality and richer than the decisions and directives issued by our leadership organs'.

Many other governments are attempting to internalize this commitment to experiment, promoted in the past by figures as varied as Francis Bacon, John Stuart Mill and the philosopher Karl Popper. The award of the 2019 Nobel Prize in Economics to Esther Duflo, Abhijit Banerjee and Michael Kremer for their 'experimental approach to alleviating global poverty' gave additional impetus to this shift, which was also promoted by, for example the Treasury Board in Canada[23] and the Prime Minister's office in Finland.[24]

I was closely involved in some of these experimental programmes, including the Innovation Growth Lab, which tested out policies to support innovation and business growth (and found that much of the money spent on these programmes was wasted) and the Behavioural Insights Team, which has run RCTs all over the world.

These methods are not panaceas, and are ill-suited to some kinds of systemic change. But they can be powerful in their effects. A good example was 'Randomized Evaluation of COVID-19 Therapy' (Recovery), set up early in the pandemic to run large, fast randomized experiments to find COVID-19 treatments. The programme quickly showed that dexamethasone cut COVID deaths by a fifth to a third, identified four other effective drugs, and confirmed that popular treatments, including hydroxychloroquine (Donald Trump's favoured solution), were useless. At least a million lives globally were saved as a result.

Here we see a governance answer to uncertainty. Instead of seeking immutable truths or long-lasting laws, governance is deliberately made more provisional, contingent and open to learning. A few years ago I suggested this is also the best way to think about the regulation of fast-changing industries, showing how regulators could create 'sandboxes' for innovators to test out ideas; testbeds; provisional licences, all with expectations that data and knowledge will be shared. These methods are now common in the regulation of finance and the UK government established a fund to help other regulators use these methods, whether for drones or artificial intelligence in law.

Since then, several academics have made similar theoretical arguments. The economist Dani Rodrik and political scientist Charles Sabel, for example, have suggested the need 'for building dynamic governance arrangements . . . under conditions of uncertainty and learning, through ongoing review and revision of objectives, instruments, and benchmarks'.[25]

9.4 How to guide what you don't understand: the principle of triangulation

So how can politics, sovereign politics, not just use scientific methods but also better control or guide science? And what are the marks of a healthy relationship between politicians and scientists, and between the public and scientists? There are many areas of life where we depend on others who know and understand much more than us. These range from the everyday tasks of plumbing, fixing cars or providing electricity in the home, to the workplace where the bosses know that the people responsible for managing money can be tempted to skim off funds, taking advantage of the financial illiteracy of their bosses, or in healthcare where the doctors may be tempted to recommend operations we don't need if that will get them paid more.

Similar considerations apply to the work of governments, which are beset by what are sometimes called 'principal-agent' challenges. These are the difficulties faced by any 'principal' (which could be a minister or the voting public) who wants to be sure that the agent, who acts in their name, truly serves their interests.[26]

We depend on specialists and we don't want to have to make decisions they are better equipped to make. We would not want surgical operations or monetary policy or the design of railways to be done by voters – people like us. Instead, we rely on methods and proxies to ensure that people who know more than us can nevertheless serve us. These methods become particularly important in science because the gaps in understanding are likely to be greater than in any other field. A scientist who has spent decades advancing a particular field of knowledge knows things that are far beyond the grasp of an intelligent generalist, including politicians, who have only a sketchy idea of the full implications of different choices on the frontiers of fusion energy or quantum computing.

So how can these imbalances be resolved? How could a parliament or a minister know which scientists to believe and what science to trust? A first set of methods involve judgements of the individuals themselves. We try to judge their trust-worthiness and bona fides, looking at their past achievements. We look them in the eye and search for signs of integrity. We look at the methods they use and whether these are reliable. This is essentially what happens when politicians and governments appoint scientific advisers.

A second set of methods draws on the judgements of others: we look to see who has the support of their academies or disciplines; who has won prizes. These are similar to the aggregated judgements, rankings and ratings used in so many other fields, often supported by the Internet. They reduce risk; but they also reward conformism and can penalize the quirky, idiosyncratic genius.

A third cluster of methods create specialized roles to make judgements on our behalf: inspectors, regulators, auditors, who confirm that the experts or providers are acting with integrity and in our interests, sometimes helped by systems of formal certification or validation that give us some comfort that we are not being deceived or robbed.

A fourth cluster of methods uses incentives to align the actions of principals and agents. For example, you can choose to reward whistle-blowers who show up fraudulent research methods. In the US the SEC offers whistleblowers ten to thirty per cent of the money it saves as a result of their actions, which often amounts to many tens of millions of dollars. Or you can choose to only fund scientific research outcomes and outputs rather than inputs (which is the method used by challenge prizes, such as the eighteenth-century Longitude Prize).

Finally, we use conversation and dialogue: engaged accountability. We ask our experts to explain their choices, particularly when things go wrong. This accountability helps to connect our world to theirs and the deeper and richer the conversations are, the more likely it is that our needs and the providers will align.

Some of these methods situate the question of trust and reliability within a context of relationships.[27] Others are much cooler – using data, track records, formal credentials and aggregated evidence. The former are more natural to us as human beings. But we learn over time to rely as much on the latter, and perhaps, to distrust our own instincts.

Together these five clusters of methods help us to thrive in a world where we depend on a multitude of experts whose knowledge we barely understand. In the case of doctors, for example, we rely on all of these. We like to believe that doctors are generally decent, altruistic people (and most of the time they are). But increasingly we can also see data about the performance of individual doctors or hospitals. We can rely on their rational methods of diagnosis and the sophisticated systems for orchestrating evidence (we are reassured if we see them reading the *Lancet*). We expect health systems to monitor them, and to strike them off if they break the rules. And, finally, we expect them to be able to explain their actions, for example in occasional public inquiries when things go wrong.

A very different example of the challenge we face in guiding technologies that are only dimly understood is artificial intelligence, as algorithms are used in billions of decisions, but with very little use of the methods described above. In 2017 the city of Rotterdam started using a machine learning algorithm (created by consulting firm Accenture) to generate a risk score to guide the city on who it should investigate for benefit fraud. The algorithm used data about past crimes, but also data on age, gender and Dutch language ability. In 2021, after an investigation by the Dutch government, the algorithm was discontinued. A media investigation found that 'the data fed into the algorithm ranges from invasive (the length of someone's last romantic relationship) and subjective (someone's ability to convince and influence others) to banal (how many times someone has emailed the city) and seemingly irrelevant (whether someone plays sports). Despite the scale of data used to calculate risk scores, it performs little better than random selection.'[28]

So here is a tool that appears to be sophisticated science but involves highly political and ethical judgements. The citizens in whose name it worked could not use methods 1, 2, 4 and 5 to judge it, but method 3, independent investigation, was the only method that could get into the black box. Accenture promised that this was a 'sophisticated data-driven approach'. But it turned out to have many built-in biases, particularly by gender and ethnicity, as have many other uses of AI. As municipalities around the world increasingly depend on algorithms, from Allegheny County in the US, which uses AI to predict children at risk, to the UK courts system, which uses algorithms to predict who will commit crimes while on probation, such investigations become even more necessary.[29]

Similar considerations apply to keeping scientists honest. But, as with AI, the methods used tend to be thin rather than thick, less extensive and less reliable than they could be. There are few available data rating or judging scientists on any measures of public interest. Scientists' training often has no place for ethics, or social impacts. There is still no equivalent to the Hippocratic oath for scientists, despite several attempts. We may have some grounds for confidence in the scientific method, but we are also aware how often those methods are flawed. There are some infrastructures and systems for validation, but these are at best uneven, whether in relation to biohazards or artificial intelligence. Finally, there are relatively few occasions when scientists have to explain their decisions and actions to a wider public.

These are all signs of a sometimes problematic relationship; and they matter because trust cannot be taken for granted. It is not enough for scientists to be seen as expert, competent and honest, or even to show 'epistemic responsibility' – a visible willingness to take on new knowledge. Rather, the relationship between science and the public is a relationship, with all the complexities that entails, and it's a relationship that often goes awry if people don't believe that it is authentic and reciprocal.

Health is an obvious example. There are many countries where significant communities don't fully trust the world of medicine. Commenting on the ways that black communities often distrust the US health system, one academic writes that 'patients and families worry with good reason that while the interests advanced by the US healthcare system might sometimes coincide with theirs, these interests are not systematically valued because they are their interests. At the same time, a patient (or parent, or family member) may decide to trust a particular doctor, nurse, clinic, or hospital, owing in part to their positive responsiveness. This medical professional, this group of medical professionals, deserves my trust not only because they are highly skilled but also because I have good reason to believe that they care about me and people like me.'[30]

These patterns are even more stark when it comes to colonial relationships, or those between dominant groups and indigenous peoples. One study described a very common pattern, in this case in northern Canada, where 'the Inuits' perception of scientific experts was strained by the fact that, normally, the environmental scientists used to arrive, collect data, and leave, without sharing what they had found'.[31]

In other words, it is not enough for the public to be well informed. Instead, science needs to make sense as part of a relationship that involves two-way communication; that is not purely extractive; and that ideally shares power, and, ideally too, mobilizes the many tools mentioned above to build confidence where there are big asymmetries of knowledge. This, in short, is our challenge. We depend on science but we have only weak methods to align it with our needs and interests and, too often, the quality of the relationship is simply ignored or taken for granted.

9.5 Democracy shaping science

If governments struggle to shape science it's not surprising that this is even harder for ordinary citizens. As I showed earlier the complex interplay between science, states, and business has tended to exclude the public and the public interest. Indeed, this is one field of public policy that has resisted democratization. Decisions have been kept under the control of scientists themselves, or of states concerned with the interests of the state itself rather than the public, or of businesses naturally concerned with business interest. The public are expected to be more observers than players, to be grateful recipients of the largesse of science, not shapers.

There are sometimes good reasons for keeping democracy at bay.

Crowds are not guaranteed to be wise in relation to speculative science, or much else that is highly technical. The more fundamental or theoretical the science, the less obvious space there is for democracy. The full democratization of science is no more desirable than the full democratization of art or monetary policy. The public lacks the time, knowledge or inclination to become deeply involved, and votes are an inefficient way to reflect complex preferences. We have little reason to trust our fellow citizens to judge wisely enough, to work hard enough or study enough to make decisions that will serve our interests.

But these are not good reasons for ignoring democracy in setting the goals of science and sometimes in shaping its applications. Indeed, it is hard to see much justification for the striking gap between what publics say they would want science to prioritize and what is actually prioritized, whether in public or private funding. In most countries funding and priority setting is captured by special interests: on the one hand, industries like aerospace and pharmaceuticals, and on the other hand the scientists themselves. That state interest trumps public interest is visible in the spending patterns of the US for example (with roughly half of federal spending still focused on the military). The gaps are very visible in design, with notorious biases in whole fields that ignored women's interests, their bodies and their needs.[32] We know much more about just how biased AI algorithms can be, particularly if they are created by a narrow, unrepresentative part of the population.

Concorde is an even more classic example of social myopia, and the gaps between technology policy and the public interest. Financed by the UK and French governments in the 1960s it was a remarkable technological achievement but one that prioritized the mobility needs of the wealthy rather than the majority (whose needs ended up being better served by the passenger plane innovations of Boeing and later Airbus), let alone the needs of others like the millions with disabilities.

There have been many modest attempts to rectify these imbalances and democratize science – using citizen juries to advise on bioethics; assemblies on climate change; and deliberations of all kinds. Denmark has a tradition of 'consensus conferences'[33] and many forms of deliberation have been tried.[34] The European Parliament has experimented with various formats – like its online consultation on the role of the chemical industry in Green transitions.[35] In the UK, an arms-length organization

Sciencewise has been funded over many years by the government to run consultations in dozens of fields, from genomic medicine to drones and nanotechnology.[36] In the US, the National Institutes of Health tried to involve the public in reviewing grants, and the US army at one point brought the public into peer review of their clinical research programmes.[37]

These experiences show that a random sample of the public, given time and briefing, can give useful steers about how to balance different perspectives. They tend to reject both the dogmatic position of scientists that rejects any rules and restraints, and the even more dogmatic position of many religions and campaigns, who want to block out new knowledge altogether.[38]

Some municipalities have also provided a closer link between democracy and science. Amsterdam was one of the first cities to appoint a chief scientist. Many US cities committed to action on climate change when the federal government was refusing to act. San Francisco passed legislation against the use of predictive analytics for policing, arguing that technologies that were not adequately understood should not be used. And, all over the world, issues such as clean air are more relevant and alive at a local than a national level: London's introduction of an ultra-low emission zone is a good example, helped by evidence about the impact of emissions on children's health, and the influence of a social movement, one of many examples where social movements played decisive roles in pushing issues onto public agendas, from climate change to mental health and waste.[39]

The institutional question then is how best to translate democratic wishes into effect. I mentioned earlier the Human Fertilization and Embryology Authority as a good example of an institution that drew on broad-ranging public debate and had powers to act, navigating the virtues of innovation and experiment on the one hand, and ethical considerations on the other. Similar institutions are already needed for many other fields of scientific advance, notably AI and synthetic biology, and more will be needed in the future.

The activities needed in any systematic approach to the governance of science and technology include:

- Analysis and exploration, ideally using a wide variety of tools from

data analytics and models to scenarios
- Assessment and interpretation, again using a variety of tools, including attention to potential social and economic impacts
- Action, linking the full range of tools that can be used, from law and regulation to funding and encouragement, and then
- Adaptation, ideally framed by explicit thinking about the potential triggers or warnings that might warrant new actions.

These are the necessary elements of effective governance, and the more they can be carried out openly, with the active engagement of elected representatives, the public and media the better. They could all be the responsibility of integrated 'technology shaping commissions', or such commissions could complement the many existing public agencies that have more silo-ed duties, typically with some responsible for funding, others for law, policy and regulation. Such commissions could have 'double key' powers in relation to the public funders of research and existing regulators, with, at a minimum, powers to ask questions, delay decisions and recommend reviews to parliaments and, at a maximum, stronger powers to speed up, slow down or direct technology to better serve the public interest.

These tasks do not need to be carried out by a single organization or group, and there are some advantages in keeping them loosely coupled, since they benefit from different mindsets. Analysis and exploration can be expansive; assessment by its nature closes in; action closes in even more and has to attend to the practicalities of action and politics; adaptation is often best organized at one remove from action, otherwise decision-makers risk being too attached to their previous decisions.

But the more these can be explicitly organized with links, with those involved in each stage required to draw on the other stages, the more intelligent the system as a whole can be.

9.6 Democratizing the priorities of science: science for society

You might expect a large part of science to be for society, or at least that a large part of publicly funded science in democracies would follow the priorities of the public. Certainly, spending on health research is high and rising, designed to drive up longevity and the quality of life, as is research on food and energy. The US National Science Foundation (NSF), which spends nearly $9 billion on research in the natural sciences, social sciences, and engineering, claims that the knowledge it produces contributes not just to security and economic growth but also 'the general welfare'. But there has been surprisingly little engagement by the public in determining what that general welfare might be, or what priorities for science should be. Social considerations, including the likely impacts on class, gender or race, tend to come low down the list of concerns – when decisions are being made.[40]

The Internet is a classic example, the result of military investment through DARPA that later fuelled an explosion of R&D within governments and business. But very little of this R&D addressed the social impacts of the Internet, the potential damage to childhood, the effects on loneliness, democracy or mental health. There is now abundant evidence that the rise of the smartphone has been associated with dramatic declines in mental health, particularly among teenagers. But no societies had strong institutions to consider these trends and their potential remedies. Twenty years later a similar pattern has been repeated as virtual realities of all kinds become a larger part of daily life.

Communications technologies have often shown up the gap between the public's priorities and those of their designers or policy-makers. Carolyn Marvin's *When Old Technologies Were New: Thinking About Electric Communication in the Late Nineteenth Century* showed how disappointed the engineers of the early telephone were when women became avid users of phones for what they saw as trivial conversation and gossip.[41] A century later the pattern was repeated when the early electronic messaging systems like Minitel, and later the Internet, were used for chat about sex, once again to the dismay of those in charge of the technologies.

In health, pharmaceuticals have generally been privileged over other fields. Drugs that had to be used repeatedly, and by patients in rich countries, were prioritized over drugs that only needed to be used once and in poor countries. This imbalance – which coincided with the long-term decline of productivity in pharmaceuticals research, eventually prompted remedial action by some governments and philanthropists like the Gates Foundation, reorienting spending to diseases like malaria and TB.

I commissioned a detailed survey in the mid 2010s of what the public (in the UK) would want to be done in their name in terms of research and technology funding. This found a striking pattern that was illuminating about the landscape of values and views. A significant minority – nearly a fifth of the population – was well disposed to innovation of any kind, including GM crops and nuclear power. They tended to be more highly educated, more urban, and more male than the average. This group seemed to include most decision-makers.

At the other end of the spectrum a smallish group – around 16 per cent – were hostile and sceptical. But the majority in between took more complex views. The largest group, about a third, recognized the need for change and innovation, but prioritized ethics and rights over innovation for its own sake, and tended to worry about waste and the effects of technology on children. For this group, who were predominantly female, innovation in fields such as care mattered more than hardware. But despite the size of the group, they had relatively little influence on policy.

To quote the report: 'There is a strong class dimension to public attitudes to innovation. People falling into the most affluent (ABC1) socio-economic grades: are more likely to be comfortable with the pace of change in society; tend to take more risks in their personal lives and see risk taking as a driver of progress; are more likely to be long-term planners; and back higher spending on research and development by business and government.' Men are more likely to support science and innovation for their own sake, women for their practical benefits.

We asked the public to rank different priorities for research. Healthcare, energy, agriculture, education, the public sector, communications and transport came top. Defence and space exploration came last, yet these are usually the priorities for states. Here we begin to see a very different possible landscape for science and innovation in the future if democracy became a more powerful guide to decisions. There are not many similar surveys. The public are regularly polled on dozens of issues but science and technology are assumed to be less relevant for public input. Yet one of the rare surveys on public attitudes in the UK (in 2022) showed that 57 per cent knew 'nothing' or 'not very much' about R&D; only 39 per cent felt R&D benefited 'people like them'; and two-thirds said they had no knowledge of R&D in their areas.

Just as surprising is the lack of methods within governments to assess the usefulness of the different areas of research they fund. Despite a strong rhetoric of impact there has been little serious research on what delivers the best returns and social value, and certainly far less on social impacts than on how science might serve the economy.

When I was in charge of strategy in the UK government, I repeatedly asked the Health Department how they decided what research to fund, and was repeatedly pushed back. In private the best answer I got was that it depended on status and relationships. Different branches of medicine would seek to install 'their' people in the primary decision-making positions and would be rewarded accordingly. Most of the government's Chief Scientific Advisers came from within what's sometimes called the 'biomedical bubble', the network of universities and institutions in south-east England that dominate the life sciences. These individuals were often impressive. But, perhaps inevitably, they worked hard to favour their own community and to block sceptical questioning or analysis. When the pandemic hit they were convincing in discussing infections and vaccines, but had little to say about the huge issues of mental health or the care sector.

A recent detailed study of the UK estimated that just 5.4 per cent of research funding is spent on prevention and public health research. More than 80 per cent goes towards biomedical research.[42] Sometimes this privileging of pharmaceuticals paid off, as when vaccines were developed with extraordinary speed in 2020. But these occasional successes masked a picture of longer-term relative failure. The data show rapidly declining returns, sometimes referred to as EROOM's law, a mirror of Moore's law, which predicted continuous improvements in digital technologies.[43] Over fifty years each billion of spending has delivered ever fewer useful drugs than in the past. By contrast, spending to influence public health behaviours can achieve high returns on investment but tends to be deprioritized. Yet the best available evidence suggests that health services account for just ten to 25 per cent of health outcomes;[44] less than other contributory factors, which include genetic inheritance, environment, social and behavioural factors. Cancer research now absorbs vast sums, fifty years after President Nixon announced a war on cancer. But even if all cancers were eliminated global life expectancy would only rise by a couple of years.

The result is historic under-investment in prevention research and preventive services. Many public health practitioners advocate using multiple interventions together, but this can be challenging to orchestrate and makes it difficult to isolate cause and effect.

In the words of one researcher, 'research funding, research activity, and the published evidence base are all heavily skewed towards studies that attempt to identify simple, often short term, individual-level health outcomes, rather than complex, multiple, upstream, population-level actions and outcomes'.[45]

Another good case is the imbalance in health research between physical and mental health. The former is massively funded, institutionalized, with strong systems for commercializing drugs and treatments, the latter is fragmented and poor. I recently commissioned a global systematic evidence review on population level mental health – what should governments do if they wish to improve the mental health of a third or more of their population who may be suffering from anxiety or depression, and not just the one or two per cent suffering from the most acute needs.

We found plenty of evidence, much of it encouraging. A lot is known about how to organize face-to-face and online therapy, how to mobilize peers or how to provide services within schools or large firms. But the scale of both action and research is an order of magnitude smaller than what's available for physical health, which is one reason why during the COVID-19 pandemic there was so little attention paid to the mental health costs of lockdowns as compared to the physical risks of infection.

In short, the organization of science is rather less scientific than one might expect. Priorities are set according to hunches, status and access as much as evidence. There is little or no public input to the task of setting priorities and politicians often don't dare to challenge high-status scientists.

9.7 Choosing pathways

What democracies need is not just to establish relative priorities. They also need to shape pathways. Technologies, and elements of scientific knowledge, rarely exist alone. They evolve in constellations, systems of interconnected elements that set a path. The car is an obvious example. Much of the world committed to a pathway that combined expanding private car ownership, investing in roads and motorways, developing the arts of traffic management, and a vast supporting network of garages, repair shops, driving schools, traffic police, supply chains and more. This pathway also meant de-prioritizing cycling or walking and organizing urban planning to maximize mobility rather than to minimize travel distances and commuting and, in some countries such as the US, de-prioritizing trains and trams.

Or take food. Very different pathways can be taken towards ever more industrialized agribusiness, mobilizing pesticides and fertilizers with automated crop picking (as in parts of the US) or alternatively towards high technology and high quality (as in the Netherlands, now the world's second most successful agricultural exporter) or again to more organic methods (as with New Zealand).

The Internet is a live example of the choices. Everywhere it rested on many other elements, including networks for physical transmission, routers, algorithms, devices, but then allowed choices between the US approach of commercial platforms, India's emerging alternative based around some open-source technologies (the 'India Stack') and China's vertically integrated alternatives, run by companies such as Tencent and Meituan, and closely overwatched by the state. These pathways are sometimes presented as fixed or given. But they are matters of choice, choices that are hugely important in part because once embarked on they are very difficult to change.

A healthy democracy should discuss its options on pathways, which requires serious analysis of the options and their implications, and then dialogue on their pros and cons. Some of the poorest countries for example now have the option of leapfrogging over carbon intensive energy sources to go direct to solar, just as a generation ago some countries jumped directly into using mobile phones for everything from payments to health.

Sometimes these are binary choices. India for example has alternative routes for developing rice seed varieties that are likely to be resilient to climate change. It can breed new seeds in laboratories or conserve and develop seeds from indigenous plant varieties. The two choices have very different interests, different mindsets and different challenges.

These are subtle, detailed tasks, not amenable to broad brush generalizations. For example, blockchain technologies bring with them both huge potential and many risks. Debate about them often polarizes between naive enthusiasm and assertive scepticism. It's true that blockchain crypto currencies can be used by organized crime networks but they can also be used to provide cheaper financial services by bypassing existing banks. It's true that they can be much more energy intensive than alternatives, but they can also be used to track provenance of materials, or feedstock, or to track pollution levels as part of systems to ensure more responsibility. It is only through detailed analysis and debate that it's possible to shape and guide pathways that in the long run make sense.

9.8 Slowing productivity and stagnation: science's social contract

The prestige and political influence of science flows from its ability to deliver. This underpins its implicit social contract. What it delivers includes economic growth, and extraordinary achievements like landing a man on the moon.

But, despite the achievements, the claim that science delivers has been tarnished in recent years. Through much of the twentieth century science successfully proved its role in driving economic productivity. The economist Robert Solow attributed the majority of productivity growth to new knowledge and its application, sometimes measured in the rather odd metric of 'total factor productivity' (which captures the causes of growth that can't be captured by other measures). The OECD and others provided extensive evidence on the returns to investment in R&D and persuaded many nations to invest heavily – Israel, South Korea, Finland committed over three per cent of GDP to research and development – all with the added motivation of achieving technological superiority over their much more powerful, and sometimes aggressive neighbours.

But in the twenty-first century an uncomfortable pattern showed up that undercut this confident claim. This was the evidence of declining productivity in R&D: more money did not easily translate into more invention or better technologies. It first became apparent in pharmaceuticals where it gained the name 'EROOMs law', mentioned earlier, the inverse of Moore's law which since the 1960s had forecast continually rising returns in IT. Instead EROOMs law showed a linear decline in the value of drugs produced for each billion of investment.

Later research suggested the problem might be more general. Studies showed a steady decline in the productivity and creativity of research. Some suggest we may be depleting a fixed resource, like the seams of coal beneath the surface of the Earth,[46] with ever fewer new ideas emerging for each dollar or euro invested. This was the argument of an influential paper by John van Reenen and others, which asked if 'ideas are getting harder to find' and answered that they were.[47]

The patterns are complex – physicists see a golden age from 1910 to 1930 that has not been repeated, whereas chemistry thrived more in the second half of the century and computer science bloomed more recently. Generalizations can be misleading. But historians increasingly conclude that the pace of progress may indeed have slowed. Compare recent decades to the decades in the late nineteenth century which brought the first transcontinental railroads in the 1860s, PVC, the telephone, phonograph and light bulb in the 1880s, the cholera vaccine, electric power stations, motorcycles, petrol powered cars, radio waves, pneumatic tyres, punchcard computers and photographic films in the 1880s, X-rays, wireless telegraph, A/C current and radioactivity in the 1890s and then the airplane, vacuum tube diode, electron and helicopter in the 1890s. Our investment in novelty and innovation is vastly greater now than during that period, but the results are perhaps less impressive.

One reason for any slowdowns is that science has become both personally and organizationally more demanding. The age of Nobel Prize winners has risen and the size of teams involved in science has grown. The growing 'burden of knowledge' makes it necessary to spend much more time understanding developments in one's field before it's possible to make new breakthroughs.[48] Studies show that the age at which academics publish their first solo-authored article in a top journal rose from thirty to thirty-five between 1950 and 2013 for mathematicians and between 1970 and 2014 for economists.[49] Perhaps, too, if Max Planck was right that science proceeds 'one funeral at a time', then rising life expectancy, and fewer funerals, should mean less progress.

There is no definitive view of whether this is a short or long-term trend. I'm doubtful that ideas are a finite resource that can be depleted. But these patterns raise questions about the relationship of science to democracy and government. Does the science system have the means to address its own problems of declining productivity? Should it be called to account for taking more and delivering less?

Most governments feel that they should have strategies for productivity since productivity determines so much else. But few have any strategies for raising productivity in science, and the beneficiaries of the existing system – the eminent scientists most likely to sit on committees or advise – are the least likely to want to reform it in fundamental ways.

Yet there has been much debate on the margins about how to reverse this declining productivity. Some focus on how research is funded and propose innovations: schemes that give money with no strings attached;[50] use lotteries or random allocations of funding; deliberately 'unbureaucratic' methods like FastGrants which aimed to dispense COVID-19 research money in 48 hours instead of months; and attempts to cut the growing proportion of time spent by researchers writing failed applications.[51] All of these have been encouraged by evidence on the capricious links between funding and impact, which show that it is much less predictably effective than anyone might wish.[52]

The management of public innovation also looks less dynamic than the management of innovation in business. The era of big corporate R&D labs is largely over. Bell Labs was at one point the world's most extraordinary creator of new knowledge. Others with striking track records of success included Du Pont and Xerox. But now the landscape is more complicated – with a much bigger role for start-ups (like BionTech, pioneer of the COVID-19 vaccine), and partnerships (like OpenAI, pioneer of the various generative AI GPTs) and much more research happening in universities.

Others hope that technology might provide answers. Machine learning, in particular, may turn out to be the 'invention of a method of invention' (according to a recent NBER paper),[53] as demonstrated by DeepMind's successful generation of new proteins with its Alphafold algorithms. A historical parallel is the invention of double-cross hybridization which made possible the development of a wide range of new corn varieties, optimized for different conditions. Eric Schmidt, former chair and CEO of Google, for example, has funded many programmes to speed up and broaden scientific innovation, which he argued is 'too often defined by new use cases for existing technologies or refining previous advancements, rather than the creation of entirely new fields of discovery', advocating the need 'to accelerate the next global scientific revolution – by supporting broad and deep incorporation of AI techniques into scientific and engineering research'.

I turn later to the imbalance between accelerating the discovery of new knowledge and the lack of any acceleration in methods for judgement or being wise about knowledge. But for now the key point I wish to emphasize is that although there is no inherent reason why the slowdown in science productivity should continue, it raises fundamental questions about the social contract between the public – who provide large sums of money for science – and the returns they get. Few politicians appear to yet understand these issues. So, although the pressure to change is likely to have to come from politicians asking searching questions, they may lack the confidence to challenge science in this way. Yet, ironically, the best answers that politics can impose on science will come from applying a more scientific approach to science itself. A scientific approach might include observation and analysis of the key facts, including the facts about productivity and impacts; experiment with different ways of organizing money; and new methods for discovery, such as AI, all fuelled by a healthy scepticism about whether the current ways of working are the best.

9.9 The public as makers of science

In the early 2000s a prominent climate sceptic organized the 'Surface Station Project'. Volunteers were recruited to show flaws in the siting of thermometers, which would in turn show climate change to be exaggerated. Instead, the evidence gathered led to a peer-reviewed article, which confirmed that the record of temperatures was robust, and that the climate was indeed warming. Here we see an interesting hybrid of science and democracy, argument and resolution. I've already suggested how democracy could remake the ends of science, and could ask demanding questions about its means and guide its ends. But this is an example of how the public can play a part in the making of science.

Citizen science has already transformed the everyday work of astronomy and zoology, counting and mapping stars and animals. Millions track numbers of birds or insects, providing a systematic map of the health of the natural world. Others look out for signs of tumours, or new planets. During the pandemic its significance was obvious in projects like Zoe, a project based in London that mobilized several million people to map their symptoms, often showing insights that were not yet captured by the formal surveys used by governments.[54] 6,000 pro bono scientists and researchers – named the US Digital Response – also helped US state governments develop COVID-19 responses, including establishing online data dashboards to track hospital resources.

Offshoots of citizen science include indigenous citizen science, which mobilizes indigenous peoples to take more charge of their own flora and fauna, and their potential uses. In health, millions of patients are organized on platforms to accelerate the discovery of new treatments, with a very obvious link between interests and engagement.

Most of the people involved in citizen science observe a strong ethos, a belief in the redemptive power of science, particularly in relation to observation and interpretation, and there are some interesting examples of when citizen science can better embody the values of science than formal institutions, like the notorious case in which the British Chiropractic Association took the writer Simon Singh to court, only for citizen bloggers to dismantle the credibility of the BCA's evidence.

These are examples of large-scale specialized knowledge rather than of democracy in a classical sense. In most citizen science projects, a power law governs how much people input, with one per cent of contributors contributing perhaps 50 per cent of all work, and ten per cent contributing 90 per cent. These patterns of contribution are at first glance very different from the principle of one person, one vote, which underlies democracy. But this pattern is in fact quite similar to real democracy, which likewise relies on the intense activity of a small minority.

Looking ahead, the work of citizen scientists points to a future where a much larger proportion of the population could play a part in discovery. Already tens of millions of people test out new diets or fitness regimes and measure the results. Within public services, teachers and police officers run rough and ready randomized control trials, to test out new methods of teaching or clearing up crimes. Communities can be challenged to cut obesity or improve the environment (my former organization Nesta, for example, ran a 'Big Green Challenge' in the late 2000s to reward the communities that could show the biggest cuts in their carbon emissions). And universities can, as a matter of course, work with local communities and businesses to solve challenges like cutting food waste or loneliness.

In all of these ways science can become a shared activity, something integral to democracy and citizenship, rather than something detached and magical. A recent study showed the potential economic benefits, which complement the social ones. Analysing over a million creators of patents in the US the study found that children born into the richest one per cent of society were ten times more likely to be inventors than those born into the bottom 50 per cent. The brightest poor children rarely become inventors because they missed out on crucial formative experiences. The most likely innovators had parents who worked in technology or were brought up in areas with strong technology industries, and so had early experience of working with gadgets, fiddling with TV sets, radios or computers. The researchers showed that US innovation could quadruple if women, minorities, and children from low-income families became inventors at the same rate as men from high-income families. But this, they argued, required shifting investment from subsidies for venture capital towards funding programmes that would expose children to innovation, and the practice of science, at an early age.[55]

10

The flawed reasoning of democracy and its remedies

> If the truth can succeed in constituting the climate and light common to governors and governed, then you can see that a time must come, a kind of utopian point in history when the empire of truth will be able to make its order reign without the decisions of an authority or the choices an administration having to intervene otherwise than as the formulation, obvious to everyone, of what is to be done.[1]
>
> Michel Foucault

It's become a commonplace that contemporary democracy is not the deliberative self-governing polity of informed free citizens envisioned by modern Enlightenment thinkers. Instead, democracy is a system of government in which public policy consists of an eclectic patchwork of half-baked programmes, where politicians tend to posture rather than act, where the public sphere is more a site of shifting amorphous moods than a clash of ideas.[2] As figures such as Yaron Ezrahi have shown, and the evidence of our own eyes confirms, real democracy is far distant from 'Imagined Democracy'. It is less about evidence and facts than about mood, perception, personality and fashion.

Even the ideal is doubtful. Theorists have questioned the – perhaps naive – hopes that the wishes of millions of people could be easily computed. Arrow's impossibility theory showed that there is no way to compute mathematically all the values of citizens and aggregate them coherently. But, even if that were possible, it might just lead to an aggregation of flawed understandings, misguided perceptions and simplistic ideas.

Meanwhile, the forms of democracy remain anachronistic. The main ones – periodic elections to send representatives to a building in a capital city – were designed for an era before telephones, let alone the Internet. Many parliaments rely on hearings that look like a courtroom and have hardly changed in a century. There is little use of visualizations or online inputs. There is over-reliance on prose reports. In countries like the

UK, extraordinarily arcane rituals and forms of address remain intact. As a result, the decisions they deliver end up being seen as neither very competent nor as legitimate.[3]

Ezrahi argues that there is 'a pressing human urge for safe-seeming, involuntary and trans-political anchorage of power'. That can be a mystical faith in majority rule, the sacredness of the constitution, or it can come from science. But if these fail, then, as Michael Ignatieff put it, their epistemological frame becomes: 'I know what I know and I follow leaders who tell me what I know . . . in the void formed when political and scientific authority contradict rather than reinforce each other . . . our "transpolitical" anchor comes to rest, precariously, on the ocean floor of our solitary selves.'

In response many have suggested reforms that might make democracy better informed, more enlightened, and less precarious. John Stuart Mill, writing before the age of universal suffrage, suggested that steps should be created from single to 'plural' votes, respectively for the unlearned and the cultured.[4] The most educated would have more votes than the least educated. This, he thought, would be better than linking votes to property, which was then the preferred model. Instead, the right to participate in public decisions should in some respect reflect the capacity to take part in those decisions. Otherwise, if equality of voice overlaid stark inequalities of understanding, the result would be worse decisions that would end up damaging the public interest. Shortly before Mill, and for similar reasons, Saint-Simon had advocated a trinity of ruling chambers, consisting of scientists, artists, engineers and captains of industry, rather like a Senate, to bring wisdom into the polity.

For a century such ideas disappeared, seen as elitist and undemocratic in spirit. But in the later twentieth century they reappeared. In the US in the 1970s an 'Institution for Scientific Judgment' (later dubbed the 'science court') was suggested to counter partisanship and the corruption of debates on difficult scientific/technical policy and legal issues. This would have been a quasi-judicial institution. It would have had active scientists as advocates, while mature scientists with diverse specialist backgrounds would take on the role of judges.[5] Its design was intended to recognize that the 'scientific and non-scientific components of a mixed decision are generally inseparable' but 'a final political decision cannot be separated from scientific information on which it must be based'.[6]

A Task Force of the Presidential Advisory Group on 'Anticipated Advances in Science and Technology' was created – and the project was openly encouraged by the presidential candidates of 1976. But momentum was lost after Jimmy Carter's election in 1976 and nothing materialized. Others suggested giving special constitutional roles to scientists. In the Irish Senate, which has weak powers, six academic senators out of sixty are chosen by limited suffrage by the graduates of just two universities.

Unfortunately, there is little hard evidence that greater education correlates with greater wisdom, commitment to the common good, justice or any other virtue. There's no doubt that the inhabitants of universities know a lot, but it doesn't follow that their judgement is superior, or that they have superior insights into the real needs of their fellow citizens. Mill's proposals for giving the educated more votes always foundered on the problem of class: that such a system simply embeds privilege. In a pure meritocracy where everyone had equal chances of success, such ideas might be more credible. But such a meritocracy exists nowhere (and in the UK for example, there is strong evidence that bright poor children have been overtaken in academic achievement by dimmer rich children by the age of ten).[7]

It doesn't follow, however, that popular will is superior to informed government. Would we rather be ruled by someone who taps into, and defers to, the shared knowledge of global collective intelligence or one who does not and relies instead on intuition, hunch and prejudice? In what circumstances would anyone wisely choose narrow over broad intelligence?

Yet it is also irrational to put too much faith in any one group of experts. So, instead, we should seek more dialectical answers – that combine popular wishes with engagement with knowledge and face up to the complexities of combining different kinds of sovereignty. In the next sections I suggest what some of the answers might be: how to separate and compartmentalize; how to grow the skills of the public and of politicians; how to organize knowledge commons that orchestrate collective intelligence to make it useful and used.

10.1 Politics protecting science from politics

There are many examples of entities designed to perform political functions which are deliberately made independent of politics on the grounds that too close a connection, and too much politicization, risks being corrupting or distorting. The principle is that the cognitive capacity of the system as a whole can be enhanced by dividing up its functions, with delegated roles for institutions that are founded on deep pools of knowledge, and that have a freedom, and even a constitutional duty, to challenge, or correct politicians and political decisions.

These institutions aim to cool – offering calm rational deliberation in place of the sometimes juddering hysteria of politics and media, and the problems of the 'permanent campaign' that have changed politics in many countries. They can counter the distorted time horizons and incentives of ruling parties, lengthening their perspectives. They can smooth out actions and reduce the risk of lurching from one policy regime to another.

Many countries have given their central banks a degree of independence, usually within parameters set by the politicians. France has various 'autorités administratives indépendantes'. The US has a long tradition of independent regulators dating back to the creation of the Interstate Commerce Commission (which was prompted by the need to regulate rail and road transport) in 1887. Sweden has a partially autonomous Public Health Agency that opted for a radically different strategy during the COVID-19 pandemic, preferring voluntary measures over coercive ones. The constitution gives it some protection from political interventions[8] (though the government can still exercise influence by cutting budgets or appointing new director generals). The UK has the Bank of England's Monetary Policy Committee, an Office for Budget Responsibility, and a Statistics Authority, all of which are independent and required to communicate to the public as well as to government.

In all mature democracies there are many similar entities, set up by law, and given a degree of freedom on the grounds that they will better serve the public interest if they are somewhat insulated from the public. We judge that they can think and act better without having to worry too much about the immediate political responses to their decisions. Their cognitive world is superior precisely because it is kept apart from mainstream politics.

But how far should we take this idea? Plato advocated rule by experts as generally superior to democracy. But few would go so far today: we want the expertise of the experts, but not to give them untrammelled power. The political theorist Robert Dahl asked a series of questions of anyone proposing to confer power on some kind of Platonian government, focusing attention on just how hard it is to make this work in practice.

He asked:

> 1. How is the Guardianship to be inaugurated? 2. Who will draw up the constitution, so to speak, and who will put it into action? 3. If Guardianship is to depend in some way on the consent of the governed and not outright coercion, how will consent be obtained? 4. In whatever way the Guardians are first selected, will they then choose their successors, like the members of a club? [. . .] Yet if the existing Guardians do not choose their successors, who will? 5. How will abusive and exploitative Guardians be discharged?[9]

One answer to these questions confers all the ultimate powers on elected parliaments, which inaugurate the guardians, devise the rules, select them and discharge any who abuse their roles. But this means that such powers can be abused by the parliaments, if they wish to pressurize or punish. Much in practice depends on norms and cultures as much as formal constitutions.

Stephen G. Breyer, before being appointed a Justice of the Supreme Court of the USA, proposed an alternative variant: the creation of a group of super-regulators who could be insulated from political pressure. They would have the 'mission of building an improved, coherent [regulatory scheme], helping to create priorities within as well as among programs; and comparing programs to determine how better to allocate resources'.[10] In other words, a bureaucratic entity would make political decisions that politics could not. Again, however, the same questions would apply: how would they be appointed, who would set their rules and who could dismiss them?

Friedrich Hayek went much further. In his later writings he proposed new assemblies that would represent people who had made a success of their lives outside politics. Strange voting systems were proposed so that at age 45 people would vote for others of the same age to be appointed for fifteen-year terms to an assembly.[11] This Legislative Assembly would sit at one remove from the more traditional Governmental Assembly, echoing Plato's rule of the wise. Other writers including Pierre Rosanvallon have suggested similar ideas to complement democracy with new types of assembly, process or role, complementing the sovereignty of the people with the sovereignty of knowledge.[12]

The critical issue for design is whether to create hybrids or parallels, synthetic models or separated ones. In one approach, these are all thought of as devices to refine the advice and inputs that pass to more traditionally elected politicians to make decisions. In others there are hybrids: committees and chambers that include both elected politicians and scientists, parliaments of both representation and knowledge. And, in the more extreme versions, new chambers – whether selected by age, by lot or by some measure of expertise, themselves make decisions.

How can we judge which approach is most appropriate for which task? The best answers have to involve some assessment of the nature of cognition in the relevant field, with a continuum from decisions that are highly technical, and only have impacts within a particular field (such as a sector of the economy, like shipping or semiconductors), to ones that are accessible to common sense and wisdom and that have effects on everyone (such as air quality).

This takes us to a very different way of thinking about democracy that's no longer a picture of absolute sovereignty but rather a shifting landscape of distinct capabilities to think and act, sometimes using arms-length agencies, sometimes relying on the public or business, sometimes exercising direct power: but always with reflexive judgements on both cognitive capacity and likely impacts. In this landscape, both the nature of the knowledge and the nature of the field matter. The more there are spillovers and wide impacts the more there needs to be some democratic oversight, even if it is indirect. But for any such decision a judgement needs to be made about the optimal cognitive conditions for decision-making: how specialized or how fast does it need to be? We can see this (for example during a crisis when decisions of great impact have to be made very fast, requiring hybrid structures and processes, and awareness that accountability will come later).

The opening up of democracy points towards a similar plurality of methods. The degrees of public involvement appropriate for decisions about a neighbourhood amenity are very different from those appropriate for monetary policy; those for intensely moral questions are different from ones that are technical; those that matter greatly to a small minority are different from those that affect the majority.

These observations take us to a view of the public sphere that is much more differentiated than the generalizations of political theory. Instead of blanket solutions, the key is an ability to deliberate about deliberation, to think about how best to think: a capacity for metacognition. Indeed, we could conclude that one test of the maturity of any political system is the variety of mechanisms it uses for different cognitive tasks, and its ability to articulate why particular methods work best for particular tasks. Devising regulations for a new field of science; regulating products on sale to consumers; monitoring potentially high-risk knowledge. These should all generate different answers that transcend either a belief in the absolute sovereignty of parliaments, or the mirror views that seek to give power to experts.

10.2 Skilled publics: shaping a public able to exercise sovereignty

Any exercise of sovereignty depends on a cognitive capacity to diagnose and prescribe, to judge and to act. Politicians need a grasp of law, policies, the machineries of government and, in Jeremy Bentham's language, the 'moral aptitude' needed for the choice of ends and the 'intellectual aptitude' needed for the choice of means. Jon Elster called this 'issue competence', the ability to choose substantively good policies, though in fact it goes far wider, since the ability to choose good policies depends on a grasp of the context, the systems in which we live, their dynamics and trends.

Elster suggested that voters only need what could be called 'voting competence, that is, the intellectual aptitude to recognize issue competence in others'.[13] But this is surely too narrow an interpretation. On many issues voters need opinions on the substance of issues, not least so that they can better judge the 'issue competence' of others who may represent them.

It follows that any society that does not consciously nurture these kinds of competence risks making mistakes, and suffering a deficit of both voting competence and issue competence. The intelligent use of science, and intelligent democratic input to science, depends on a well-informed public and, although there are bound to be asymmetries of knowledge and understanding between the public and the scientists and, although these asymmetries may be bound to grow over time, attention to this cognitive gap has to be part of any programme to better align politics and science.

In previous decades this was seen as a problem of science literacy. In an influential book Jon D. Miller broke this down into four elements: (a) knowledge of the basic textbook facts of science, (b) an understanding of methods such as probability reasoning and experimental design, (c) an appreciation of the positive outcomes of science and technology for society, and (d) the rejection of 'superstition'.[14]

This approach implied that the priority was education and explanation – through schools, television and radio, exhibitions and more. The aim was that a good proportion of the public should have some understanding of how electricity works, the nature of an atomic bomb or how infectious diseases spread. Success could be judged by large-scale surveys.

There is no doubt that levels of knowledge and understanding vary greatly. For example, a survey in the mid 2010s in the US found big gaps between public attitudes and those of scientists: in relation to 'opinions on whether humans have evolved over time, whether genetically modified foods are safe to eat and whether climate change is due to human activity, a 30-point difference or more was observed between scientists and the public'.[15] 2020 was the first year in which more than 50 per cent of Americans accepted Darwin's theory, a sign of the scale of the challenge, as is the finding that 44 per cent don't believe that human activity causes climate change.

Politicians struggle to know when they should engage with misinformation – or whether by doing so they give it oxygen. In 2014, New Zealand's Prime Minister, John Key, felt he had to announce officially that he was not a shape-shifting reptilian alien (in response to a citizen's Open Information Act request asking for 'any evidence to disprove' that he was an 'alien ushering humanity towards enslavement'), though of course if he had been an alien this is exactly what he would have done.

Some literacy is essential for functioning in a complex society. We need significant proportions of the public to be able to judge not just wild conspiracy theories but also media articles that claim 'research shows x', or to spot misinformation and deceit. We need them to recognize that 'science frames' can be more influential than 'science facts': how an issue is framed shapes what we see or argue about.[16] And we need them to see when they are being manipulated: a recent study found that the prevalence of 'emotional payload' in media headlines went up dramatically after 2010: 'each additional negative word [in a headline] increased the click-through rate by 2.3 per cent'. We may enjoy being stimulated – but it's vital that we see how it's done as social media spread opinions without context or qualification.

Scientists have long appreciated the need to cultivate understanding. In the UK the Royal Society, and later the Royal Institution, whose first two heads were Humphrey Davy and Michael Faraday, both exceptional scientists, set up weekly discourses and lectures to educate the public. Many of the greatest scientists worked hard to communicate – from Albert Einstein and Marie Curie to Stephen Hawking.

But passive receipt of knowledge is not enough. The 'public understanding of science' approach that became popular in the 1980s assumed that if only the public knew more about science they would be more supportive of it.[17] But since then the debate about science literacy has evolved, trying to escape from the assumption that it was always the public, rather than the scientists, who were the problem. The alternatives aim to involve the public not just as observers and passive recipients of scientific knowledge but also as makers and shapers. This 'relational turn' is based on the premise that science needs to prioritize not just explaining the complexities of knowledge, but also a more two-way approach to communication that involves listening, responding and opening up to the democratic input of citizens.

This task starts in schools but with an approach to teaching science that encourages engagement with the current frontiers of science that are being reported in the media, and an ability to ask questions rather than just repeating accepted knowledge. Here, changing approaches to science connect with changing approaches to the role of education in democracy, which isn't just about learning the facts of how decisions are made but also involves gaining experience in the exercise of power and responsibility.

Classroom exercises that focus on misinformation and disinformation are useful (as in Finland) and can help children learn not just how to think but also how to avoid being drawn into social media rabbit-holes (as William James commented, 'the art of being wise is the art of knowing what to ignore'.) But they can also be enhanced by experience of influencing budgets (as in Paris' participatory budgets, which allocate a portion to schoolchildren), or project-based learning in which children work on practical projects in their neighbourhoods on topics such as air quality, connecting scientific measurement and analysis with attention to issues of behaviour or economics. The more that learning is combined with action the easier it is to learn. Here, the potential for new technological tools to help us all to express ourselves democratically is intriguing: in the near future it is plausible that we will be able to brief personal AI agents to express our preferences, provide us with information, and influence debates on our behalf.[18]

A more active and engaged public can then ask the media to do a better job of explanation. There have been many initiatives to discourage the worst features of coverage, the hyperventilating claims that research shows why a particular food is either the key to health or a source of cancer. The media naturally try to dramatize, with risks more dramatic than opportunities, certainty more engaging than nuance and caveats.

Many countries have organizations that advise the media on how to communicate science. Here the details of how knowledge and risk are presented become all important. There is now good evidence that people can quite easily learn about biases, statistical literacy and decision heuristics and that this then improves their ability to make decisions, whether personal ones or collective ones.[19] If you learn about how easily your own mind can be misled by false information you become better at spotting it.[20] Similarly if you learn about dangerous collective behaviours you can avoid becoming part of them – something crowd experts call the 'faster-is-slower' effect. Learning that when a crowd rushes towards an exit the total time required to get out is longer helps a group avoid that mistake in the future.

Explaining risks and uncertainties in ways that acknowledge ambiguities without becoming too complex isn't easy. The IPCC, for example, concludes that 'Global warming is *likely* to reach 1.5°C between 2030 and 2052 if it continues to increase at the current rate (*high confidence*).' The terms 'likely' and 'high confidence' are shown in italics and 'likely' is defined in a footnote on a different page as corresponding to 66 to

100 per cent probability. Studies have shown that people interpret these words differently[21] but also that this problem can be reduced by presenting the verbal and numerical expressions together (e.g. presenting it as 'likely (66 to 100 per cent probability)').[22]

Ecological issues can be particularly complex. The more we know the less confidence we can have in our gut reactions. For good reasons global public opinion has become much more aware of the risks of plastics – to wildlife in the oceans, and to our own health. Many countries have brought in bans and charges for plastic bags and other items. But the impacts of these changes aren't intuitively obvious. For example, replacing a disposable plastic bag with an organic cotton one only makes sense in terms of carbon emissions if you use the bag at least 150 times.[23] Plastic packaging of food can often be better than the alternatives once you take into account the resulting degradation and waste. A global ban on plastic straws would achieve at most a 0.03 per cent reduction of plastics in the ocean. In these and many other examples the detail is all-important; science can help us to understand, but also risks prompting the response that it's all too difficult.

Here again participation in science may be crucial for deepening understanding. The many citizen science projects mentioned earlier such as Zooniverse and Foldit have mobilized millions.[24] There is reasonably strong evidence that direct participation in such projects increases understanding and appreciation of science, and there is a more general cognitive bias to think favourably about things you have played a part in creating. In marketing there is evidence that if a product is presented as user designed or crowd-funded it is seen more positively, a process of 'vicarious empowerment'.[25] This suggests that the more science is presented as a collective creation, involving many, the more it may be trusted.[26] It also suggests that the data presented earlier, which showed that much of the UK public know nothing of the R&D done in their locality, could be a signal of future problems.

Greater engagement and understanding may also encourage more sensitivity to uncertainty and ambiguity. Anyone who has taken part in experiments soon gets a feel for the complexities of sampling and inference. Some of the population may still cling to certainty and simplicity: perhaps a third in many countries are disposed against complexity and they form the political base for authoritarian leaders.[27] But others are more flexible.[28] Recent research suggests that dogmatic, inflexible thinking is a kind of metacognitive failure at the individual level and that it can be mapped, and is associated with over-confidence, and 'a generic resistance to recognizing and revising incorrect beliefs'.[29] This research also suggests that young people can be taught to avoid these deficits of metacognition. Here, perhaps, we see the outlines of a more authentically twenty-first-century approach to education, cultivating better skills at metacognition at the individual level that can then support metacognition at the societal level too.

A more aware population makes it easier for scientists to combine clarity with honesty about limits. As one author put it 'the double ethical bind for communicating science to the public, then, is for the scientist to find an appropriate balance between being an effective agent for change and being honest about the limitations of the state of knowledge'.[30] Indeed, this is perhaps the key. We need both scientists and citizens to be good at scepticism and critical thinking, taking seriously the injunction *nullius in verba*.

10.3 Skilled politics: the case for academies for politicians

Aristotle, whose ideas on politics I began the book with, believed that a good polis needed the best people to be in power, by which he meant those with the greatest virtue, though virtue wasn't the only quality they needed. Sometimes, if different qualities were held by different groups, then power needed to be shared between them.

But what counts as virtue? And what are the qualities needed to govern a society in which science is prevalent? One possible answer was given at an event in 2016, when Justin Trudeau, who was then Prime Minister of Canada, faced questions from journalists. One of the journalists began like this: 'I was going to ask you to explain quantum computing but . . . haha . . .'. To the surprise of the audience, Trudeau responded with a sensible, competent explanation of how quantum computing works. His answer went viral and burnished his reputation. It didn't reveal a deep knowledge of theoretical physics but it did reveal an ability to read a briefing.

No one expects politicians to have a deep expertise in science. But we can expect them to have some knowledge. Indeed, if politics is to remain powerful, albeit with its power more open and shared, then the knowledge and quality of politicians, as well as the quality of the processes they use, become crucial. Here we immediately face a paradox. This, perhaps most important, of roles is treated more casually than perhaps any other. Every other position of leadership requires skills: years of education are required for lawyers, doctors and even business Chief Executives. Yet for politicians there is no training and the primary methods of selection test for qualities that only loosely align with those needed once in post.

There are a few exceptions. China invests heavily in training its leaders, who must attend party schools, write essays that are marked, and stay on top of the leading ideas in technology or law (while also confirming their familiarity with Marxism–Leninism). The US businessman and former Mayor of New York, Michael Bloomberg, has financed extensive training programmes for mayors, while in Australia the McKinnon Institute is developing an academy for politicians, with a sophisticated curriculum.[31] These can help them stay abreast of geopolitics, technology or law, or put them through simulations to understand how to make decisions under pressure or to role-play with budget allocations. But, in most democracies, it is felt inappropriate to spend public money on preparing politicians.

The result is a difficulty. Politics needs to come to positions about science. But it cannot understand what it is governing in any detail and the gulf between what it knows and what could be known grows with each decade. Science becomes incomprehensible to other scientists and even more so to outsiders. The scientists may be ignorant of politics and lack much grasp of what matters to their fellow citizens. But the ignorance of politicians in respect to science must be greater simply because the scale of the commons of science is so much larger.

It can help to have leaders – like Angela Merkel or Margaret Thatcher – with some grounding in science. China has increasingly filled the upper echelons of its leadership with scientists. But Merkel's achievements in quantum chemistry and Margaret Thatcher's in X-ray crystallography may have given them little feel for computer science or physics, any more than speaking one foreign language necessarily gives you privileged access to another one.

The result is that even the best politicians are in effect 'structurally incompetent'. I remember, for example, the challenges faced by my one-time boss Tony Blair. He was well intentioned; instinctively pro-science and pro-technology; and knew he should be engaged. But he simply lacked the experience to know what good looked like and it was easy for the charismatic bosses of big digital companies to take advantage of him.

Many struggle. But devoting a small proportion of their time to learning – with sessions to acclimatize to the frontiers of key fields of science and technology and their dilemmas – would be useful in every way for democracy. Then, in their everyday work, politicians need support with systematic ways to present the available evidence, the balance of opinions among experts, or to visualize complex issues rather than relying solely on text. The best examples separate out diagnosis and establishing the facts from generating options and then deciding between alternative options.

The aim should be to make politics and democracy cognitively 'thick': rich in shared knowledge, knowledge about knowledge and the ability to synthesize across multiple domains. Indeed, a mature polity needs a large repertoire of different ways of mobilizing knowledge – from expert panels and reviews to inquiries, crowdsourcing and citizen assemblies, fit for different tasks, and able to loop back to explain why a particular method or knowledge source is used for a particular issue.

A century ago, Oxford University created a new course – Politics, Philosophy and Economics – designed to nurture a cadre of leaders. Many British prime ministers and ministers took it (as did I). But its emphasis on macroeconomics and analytical philosophy doesn't sit well with a world of pandemics and climate change. Instead, we need a new curriculum for power: one that includes awareness of data, systems and complexity, of engineering as well as science, psychology as well as philosophy. Perhaps, too, we need something closer to a Hippocratic oath so that politicians pledge to 'seek, absorb and act on the best available knowledge', a commitment mirrored by scientists committing to 'share, open up, accept criticism and accept judgements made by others about the common good'.

10.4 Knowledge commons, superpolitics and science assemblies

The crucial bridging institutions – that can enhance the practical intelligence of both citizens and their representatives – can loosely be called 'knowledge commons'. These are the ways in which the best available knowledge is brought together, curated, analysed and evolved to support argument and debate, to enrich and thicken the cognitive environment in which decisions are made.

The quote from Michel Foucault at the beginning of this chapter was written with typically ironic detachment. But, as utopias go, one where the truth constitutes 'the climate and light common to governors and governed' is an attractive one. A shared picture of the facts, the possibilities, the choices, and a shared dialogue about their implications – an orchestrated collective intelligence in other words – is a necessary condition for wise actions, and vital for guiding science.

There are many forerunners. Health and medicine are the most advanced in terms of shared knowledge, helped by organizations like the Cochrane Collaboration and the Mayo Clinic, which synthesize medical evidence, or NICE in the UK, which judges the effectiveness and cost-effectiveness of different treatments. There are close links into doctors' training, and a clutch of online sources like PubMed. Patients have access to the same knowledge as their doctors even if they struggle to make sense of it.

The IPCC plays part of this role around climate change, with shared data and models to forecast the dynamics, alongside hundreds of websites, repositories, search tools and roadmaps to help organizations and individuals make decisions about how they can reduce their contribution to warming.[32] These are pointers to a future in which knowledge, and knowledge about knowledge, is open, shared and well organized.

The structure of a knowledge commons for decision-makers needs to encompass at least four major strands:

- **Facts:** a strand of diagnosis that covers the key facts or patterns, the dynamics of change.
- **Evidence:** a strand showing what are the proven options, which clearly work in at least some contexts and so might warrant adoption or adaptation, as well as knowledge about harms and risks.

- **Innovations:** a strand covering innovations, the promising practices in the present, whether in the field itself or other ones, that could be useful or relevant in the future.
- **Possibilities:** a strand mapping what might be possible, perhaps a decade or generation from now, that might shape choices in the present. This can include foresight, scenarios and speculations.

These four dimensions need to be kept in mind simultaneously, since all are relevant to choices and actions in the present. Each of these can be thought of as a landscape. They can be mapped visually as well as described in prose, modelled and simulated. They can be put on the walls of rooms or on screens, to help groups absorb their messages. Spending time immersed in each of their dimensions, and arguing about the messages, helps any group to better understand its options, and the dynamics of the systems they are trying to influence.

But this kind of decision support landscape doesn't emerge automatically. Google search and GPTs do not come close. Instead, these need to be carefully curated, made accessible and relevant to the choices with which decision-makers are grappling. Here, the problem is that the task of doing this falls between different institutions – governments, parliaments, the media, academia, civil society. None owns the task sufficiently. Yet the cost of organizing such knowledge commons – in fields such as AI, energy or bio-risk – is a necessary overhead for a functioning democracy.

These then provide the infrastructure to support the public argument and deliberation that is essential to support legitimate decisions. One option is to organize this more systematically through what could be called 'science and technology assemblies' that combine some of the methods of citizen assemblies to provide an evaluation and commentary on important fields of science. France's citizen assembly on the end of life, established by Emmanuel Macron, is a good example.

Such assemblies should be representative of the broader public, like juries or existing assemblies. They should be briefed by experts, and helped to understand the four dimensions mentioned above. They should be supported as they consider and reason, and then turn to potential measures to accelerate, slow down or stop. In doing so they can aim to be consensual but they can also use argument. Zeynep Pamuk for example suggests 'science courts' that use adversarial argument to flush out different options and views.[33] This is also space where science, and emerging AI, has a lot to contribute: providing syntheses of available knowledge; shaping it to different cognitive styles; helping groups to deliberate and guiding them towards consensus.

To be effective, such assemblies need to evolve beyond the current models. A crucial design task is to ensure that the deliberations of an assembly reach beyond those directly involved to the wider public, whether through the engagement of the media or recruiting people with demonstrable influence on others rather than purely random sortition. Another crucial design change is to ensure that proposals are workable. This requires close involvement of both bureaucrats and politicians in their later stages. Otherwise the risk is that, as with Macron's climate-change assembly, the recommendations will be dismissed as unworkable.

Many of the most exciting fields of scientific advance contain within them both great promise and great risk, and badly need this sort of attention. Quantum computing is a good example. Large sums are being invested[34] and it is likely that there will be big advances in medical imaging, sensing (for example of earthquakes), cryptography and complex calculations. But no country has managed to weigh up the benefits and the risks, or to set out plausible pathways for maximizing the former and minimizing the latter. Similarly, synthetic biology and chimeras could offer new ways to create organs for humans but also synthetic viruses. Cognitive enhancement technologies could improve our ability to think and remember but could also fuel pathologies, people engineered with superpowers but deliberately shaped to lack conscience or fear. There are plenty of antecedents for the more reflective deliberation suggested above, particularly in bioethics, and most of these proved wise in their judgements. In one option, such assemblies could include elected politicians as part of the deliberations. At some point politics has to become involved, in the narrower sense of the argument and compromise of current values and interests.[35] But such assemblies can also consciously take on a super-political role, considering the perspective of future generations, of the biosphere and of our duty to leave behind the best possible knowledge to help our successors.

10.5 Knowledge commons for metacognition

The most important knowledge commons of all is the knowledge commons that orchestrates knowledge about how to make good decisions. This is collective intelligence about collective intelligence and, like the other knowledge commons, gathers many disciplines, data, experiments and insights. Such societal metacognition helps to guide us on what we know and don't know; how to model; how to use visualizations; how to divide up roles; how to interrogate; how to argue; how to synthesize. Sometimes this is described as the science of science, and there is a growing field of study that gathers data and looks for patterns, for example on the problems of productivity described earlier and their possible remedies. But what's needed goes well beyond what is currently called the science of science.

In Chapter 13, I describe the science of wisdom and how its essence is a willingness to learn constantly. It's ironic that science uses relatively little science to guide its own work. As one small example: I once interviewed leading figures in many top universities to ask what science of meetings they used to design the innumerable conferences, seminars and symposia they run. None could give much of an answer: essentially they do what they've always done.[36]

It is equally ironic that governments are so uninterested in the science of how they should work, only sporadically looking at how they could better organize their structures and processes. As far as I am aware no parliaments have even small teams charged with synthesizing knowledge about how parliaments could work better and only a handful of governments have teams tasked with keeping them abreast of the best available science of government. Yet these loops are the essence of wisdom, vital for understanding and shaping complex issues that are full of ambiguity and uncertainty.

PART V

The Problem of Scales

Borderless Science in a World of Borders

11

The clash between global and national interest

During the COVID-19 pandemic, France's President Macron visited Russia's President Putin at the Kremlin in Moscow. A famous photo shows them at opposite ends of a strangely elongated table. The reason for their seating arrangement was that the Russian government had asked that Macron should take a COVID test. He had already taken one in France, but they insisted on their own test. He refused, apparently advised by his intelligence agencies that if Russia got hold of his DNA this could be used to spot weaknesses, such as vulnerabilities to sickness or that it could even be used to design targeted weapons tailored to his genetic make-up. During a period when many had apparently been murdered on Putin's orders such profound mistrust, and the fear that science could shape new political weapons, was understandable.

11.1 The evolutionary dynamics of competition and cooperation

In the previous chapters I mainly focused on the relationships between science and politics at the national level. But science and knowledge spread and sprawl, and they bring with them new types of competition and conflict as well as cooperation. Here I want to suggest that these dynamics are best understood in terms of multi-level evolution. What confers evolutionary advantage on a nation or company may be at odds with the evolutionary imperatives of the world as a whole. What helps a nation survive may threaten the survival of the species.

To grasp this, it's useful to take a step back and reflect on why science has been able to evolve at all. The simple answer is that systems of knowledge – and the logics they form part of – spread and grow if they generate positive feedback loops. These can be both objective and subjective. They are objective in the sense of achieving results for social groups and fields that hold power, money and esteem – as science certainly did by conferring military or export advantage. And the feedback loops are also subjective in that these results are recognized and appreciated.

So, when engineering and science delivered success in warfare or the economy, glory or public benefits, and when this was recognized, rewards came to science and scientists: new funding, honours and positions of responsibility followed. Within business those firms that mastered how to mobilize science for new products simply outcompeted others, and that message was understood. Jurgen Renn's work on the evolution of knowledge helpfully details the evolutionary patterns that shaped the many different forms of knowledge, from writing and accounts to laboratories and new materials, emphasizing how they adapted to better fit environments.

That history includes many methods to find useful knowledge. Trial and error is the most common – trying new food sources, tools or cures until you find one that is good enough. Another is introspection – looking inside ourselves to find clues to the world we are in, an approach which guided much philosophy and still does. Then there is the search for the specificities of things rather than what is generalizable (and various writers have imagined an alternative trajectory for science that concentrated on these, the logics of botany and zoology rather than physics). Yet another is the search for arcane knowledge, which is deliberately kept secret, for fear it might be misused, or conversely the idea of a science more rooted in lived experience, an idea promoted by figures like Francisco Varela.[1]

But the methods of modern science – seeking generalizable theories, particularly through experiment – outcompeted these other methods, offering superior observation, superior interpretation and a clearer link into action and making. That gave science, and its logic, a more dynamic metabolism that was able to draw in money, power and commitment. Much of the money and power came from states, while the commitment mainly came from people who devote their lives, hearts and souls.

The dynamics of this can be analysed both within nations and at a global scale in terms of competition with other contenders for resources and status – business, arts, governments and religions. All are always both cooperating and competing for scarce attention, love, admiration and money. Science tends to thrive when it creates value for others (states and businesses); when it's supported by widespread trust in the bona fides of scientists themselves; trust in their methods of verification;

and a rough alignment of the folk stories of science with the reality – for example the kinds of story that tell us of saintly geniuses inventing penicillin or theorizing gravity (stories that tend to be closer to much older models of thinking about magicians, achieving things through force of brilliance and will, rather than the mundane, collective processes of real science).

In global competition it was obvious that science-based nations could win wars: superior rockets, tanks and missiles could prove decisive. Later it became obvious that science-based economies – such as the US, Germany, Japan and later South Korea or Sweden – outcompeted others solely reliant on exports of raw materials – which is why so many countries deemed it wise to spend a rising share of GDP on science.

But the evolutionary feedback is blunt not acute. One definition of power is the ability to get away with mistakes, and many states were able to get away with big mistakes in technology so long as they had other strengths. Russia's science system was extraordinarily expensive, failed to win wars (Afghanistan or Ukraine), and fed an economy that still largely relied on exporting crude commodities not high-tech artefacts. Yet it nevertheless retained its status and much of its command of resources.

This evolutionary model becomes more complex if we look at the relationships between different layers. The emergence of a global science system can be understood as a new form of cooperation at larger scale, that has grown because of the advantages it conferred on nations, governments, businesses and individual scientists, and the positive feedback loops it engendered (again, with money, power and commitment flowing into it). Very few countries can shape global science: most can directly control only a tiny percentage of the total and, even if they impose restrictions, others may not. So logic points to ever more pooling of research, regulations and standards.

However, at each level the successes of science brought it into conflict with lower tier groups – national, religious, corporate – many of which use science but are also sometimes challenged by science. New agricultural methods, methods of counting or measuring, public health measures, logics of generalization, standardization, universalization: all are experienced as an imposition from above. They challenge local knowledge and authority as well as bonds, loyalties, commitments that outweigh those of the science community. The tensions are very similar to those between universal religions – Catholicism or Islam – and local power. These conflicts are fractal, emerging at multiple levels.

Local customs or knowledge may well survive an encounter with science – but they never again look quite so obvious or natural. So, it makes sense to resist, to question and to mock, and within democracies the feedback loops for politicians may encourage them to attack science – as alien, foreign, cold, imposed.

For rather different reasons it makes sense for an interest group to sow seeds of doubt, and to make science an enemy in other, subtler ways. For the tobacco industry, oil and gas, it was never science in general that was the enemy but rather particular parts of science which could be attacked with science's own tools of doubt and scepticism: deploying the standard argument that more research was needed before any definitive conclusions could be drawn.

This social reality – of contest, argument and ambiguity – sits above the apparent calm logic of accumulating knowledge. And, because humans are essentially social creatures, what we believe depends on who we believe, and what we trust depends on who we trust.

11.2 Global imbalances and the struggle between hope and fear

These evolutionary dynamics become clearer if we look at science from a global perspective. Modern science can trace ancestors, precursors of all kinds, not least in the boisterous scientific traditions of China, India and the Arab world, or the philosophizing of ancient Greece. Human history is inseparable from the history of its organization of knowledge, whether embedded in roles – like those of the priest, builder or farmer – or codified and formalized in arithmetic, scripts, theories, calendars like those of the Mayas, and disciplines like mechanics, geometry and optics where science and engineering fuse together.

But, although many prefer now to play down the extent to which modern science was a sharp break with the past, the data is clear. The 'great acceleration' that began in the late eighteenth century, in part because of the institutionalization of science, and which led to longer lives, more prosperity, more production, more energy use, more population, more pollution, more deforestation, was undoubtedly a great break, indeed the greatest break in human history.

It happened in Europe, initially in Britain, and quickly conferred an extraordinary dynamic of evolutionary feedback as superior armaments, and superior organizational methods, enabled the conquest of large swathes of the planet (for example, the East India Company's capture of northern India, then the world's richest area, which was responsible for a quarter of global GDP, ten times more than England at this time).

It's why the great civilizations of India and China that found themselves suddenly colonized and bullied struggled to understand how the terms of competition could have shifted so much and felt it as something quite abrupt and is why they have proven so determined in the twentieth and twenty-first centuries to rectify that imbalance.

Part of the relative success of Europe came from a superior ability to use global science. Newton, for example, 'relied on data collected from French astronomers travelling aboard slave ships, as well as East India Company officers trading in China'[2] and the work of later scientists was closely tied to empire. The extraordinary collaboration of many French and British scientists to measure the distance of the Earth from the sun in the 1760s, building on plans designed a century before by Edmond Halley to take advantage of a very rare moment when Venus crossed the sun, was a remarkable demonstration both of the long-termism of science and of its capacity for global cooperation.

By contrast, Ming China thought it had little to learn from others. The first approach was rewarded; the second punished. Again, the feedback loops promoted the more open, networked (and often imperialist) approaches of the West over the more closed approaches of the East.

The lessons are obvious, and explain why contemporary China works so hard to keep its scientists linked into global networks, and why even north Korea invests so heavily in grabbing superior technology by any means to maintain its position in nuclear (the USSR provided its first research reactor in the mid 1960s and it's now estimated to have several dozen nuclear bombs) and missiles (adapting Scud missiles first bought from Egypt in the 1970s).

There has been a very healthy interest in recent years in what was forgotten or suppressed from older bodies of knowledge, and the importance of respecting what is – in a sweeping generalization – sometimes labelled 'indigenous knowledge'.[3] But true autarchy is no longer an option for any community that seeks to have some control over its own destiny.

This takes us to one of the great questions of science governance: will science in the future primarily be a tool for geopolitical competition or will it be an alternative to it? Will evolutionary dynamics promote new forms of global governance – or the opposite? Will behaviour be shaped more by hope, or by fear that other nations will use power to dominate?

Ever since the nineteenth century, there has been hope that somehow science would usher in a new kind of governance – rational, benign and global. Politics would metamorphose from the national to the global, shedding its fantasies and follies along the way. In the 1940s, for example, US Vice President Wallace saw modern science as the means of giving everyone in the world enough to eat and echoed an outward looking phase of US social science, which believed it could contribute to global development. The arrival of the United Nations gave new momentum to these ideas, and the sheer horror of nuclear war appeared to justify the argument that only global government could save humanity from self-destruction.

George Orwell memorably mocked H.G. Wells about just this hope that science and war were opposites: 'On the one side', he wrote, 'science, order, progress, internationalism, aeroplanes, steel, concrete, hygiene, and on the other side war, nationalism, religion, monarchy, peasants, Greek professors, poets, horses' (a list I have more than once sent to friends from Greece who happen to be professors).

It was not obvious which side would prevail. Darwinian theories of evolution had encouraged previous generations of strategists to think in terms of a ruthless competition for survival and these ideas returned after the Second World War. In his influential book *Scientific Man versus Power Politics*, the US thinker Hans Morgenthau argued against the rationalist optimists. All global institutions were part of constant political struggle. International law was an illusion in a world shaped by the unavoidable realities of war and competition: 'the choice is not between legality and illegality but between political wisdom and political stupidity'. Statesmen were better placed than lawyers to understand this. Cooperation was a possible product of world order but there was no inevitability, indeed the odds were stacked against it. What came to be called 'realism' interposed itself, implying that science should be mobilized behind national not global interests, used in the service of power and not against it.

The role of espionage confirmed the realist view, offering an alternative route to the global spread of knowledge than the republic of science. Espionage enabled the USSR to build its plutonium bomb in 1949, and to copy US Boeing B29 bombers, part of a broader strategy of 'autarchic development based on foreign technology'.[4] China invested heavily in industrial-scale espionage in this century for similar reasons. For each of them, the nation was paramount, and the world was a zero-sum game in which the most valuable knowledge would not be freely shared.

Many reached similar conclusions from very different starting points. Edward Meade Earle, another influential figure in the new field of security studies in the middle of the twentieth century, emphasized that the risks which science was producing required a stronger national response. The theologian Reinhold Niebuhr made parallel arguments, explaining why global institutions and the end of sovereignty were implausible because of the will to power, which was bound to be mainly expressed through nations.[5] The latter, he argued, would use scientific risk as a weapon – with new poisons, bombs and missiles that had to be resisted.[6]

Many scientists worked on the alternative vision of a transnational republic, above and apart from the messy world of national politics. For example, in September 1968, while the world was in turmoil from Prague to Paris and Vietnam to China, 62 physicists gathered in Geneva to found the European Physical Society, including official representatives of the national physical societies of eighteen countries in both Eastern and Western Europe, who signed the constitution in spite of the political divides of the Cold War. The main proponent of the society, Italian physicist Gilberto Bernardini, saw it as realizing a dream: a single community of European physicists representing a culturally unified Europe that he described as a 'single highly civilized nation'.

CERN became a striking example of cooperation that went beyond Europe – with over twenty countries supporting its work to discover the fundamental particles of the universe, and occasional offshoots like the invention of the World Wide Web. Half a century later the collaborations funded by the EU's Horizon Programme have made that kind of cooperation a mundane reality, as universities from Bulgaria and Portugal to Denmark and Italy collaborate as a matter of course. Global science is strongly institutionalized too – through conferences, societies and a lattice of collaborations, and for individual scientists being cut off from global science is damaging. The Cold War, for example, separated the worlds of maths: crucial work such as Andrei Kolmogorov's on probability was scarcely known in the West. This also stunted some fields of technological development: for example, the phage therapies that were widely used in the USSR and eastern Europe were shunned in the West, until they returned in the twenty-first century as a possible answer to rising resistance to antibiotics.

The world's most successful nations – whether measured by economic indicators or indicators of wellbeing – are ones that have learned how to share and cooperate. But the larger ones at least have learned a larger repertoire that includes intensive competition, secrecy and deception as well.

It is a commonplace that global science is organized as a commons, resisting boundaries, constraints and often ownership, so that there is no British science, Chinese science or Russian science, only global science. But the story of globalized science can be misleading. There have been truly global initiatives like the Human Genome Project. Geneticists in the US, UK, China, Germany, Japan and France worked together and showed that human genetics were almost identical. But the HGP was based on a few anonymous blood donors from the city of Buffalo in the USA. Afterwards multiple alternative genome projects began, including the 100,000 UK Genome Project, the Asian Genome Project, the Chinese Million Genomes, the African Genome Sequence Variation project, as well as whole-genome sequence population studies in the Netherlands, Qatar, Turkey and Japan, and the Genome Russia Project, each investigating the variations in the human genome.[7] The Human Pangenomic Project now has 47 'reference' genomes from across the world, and helps to show that although we all share around 99.6 per cent of our DNA the differences matter, for example in relation to lactose intolerance. We appear, in other words, to be in an era when science is simultaneously global and national, universal and contextual, collaborative and competitive.

Climate change shows this all too clearly. Climate change was first theorized in the late nineteenth century and discussed by political leaders as varied as Lyndon Johnson in the 1960s and Margaret Thatcher in the 1980s. Historians now suggest that scientists in the Hapsburg Empire, such as Julius Hann, were among the first to understand weather and climate as phenomena that crossed borders and connected to global patterns, and so needed novel ways of both understanding and responding to them.[8] The scientific imperative of better understanding chimed with the political imperatives of a multi-cultural polity.

In the late twentieth century too, scientists pioneered the understanding of climate as a global system and became sentinels, warning of disaster and advocating global responses. Climate change is perhaps the ultimate example of what economists called externalities – the costs of one person's action that fall on another – and of a phenomenon at odds with the very ideas of sovereignty and borders, the constitution of the political.

Indeed, scientists showed that the realists' view of the world was not always realistic in a world where knowledge leaks and spreads, producing 'spillovers' in the language of economics, and where collapsing systems make a mockery of the economist's idea of a stable, 'optimized' equilibrium.[9]

Some, however, see climate change solely through a prism of competition. I remember attending events in China in the early 2000s and hearing senior figures from military intelligence denounce the idea of climate change as a Western plot, designed to block China's rise by preventing it from building an economy of carbon, cars and planes.

Not long after, some of the security elite in the US adopted a mirror view: having opposed climate action as a threat to US competitiveness, they switched, seeing a rapid shift to renewables and Green technologies as a way to save America from dependence on an unstable Middle East.

A few years later the discovery of vast new reserves of oil and gas within the US changed the calculation yet again. In many countries politicians have presented action on climate change as an unacceptable constraint on national sovereignty and competitiveness: Australia's leaders in the 2010s were a particularly stark example of this.

11.3 'Changes not seen in a century'

Writers in China sometimes describe epochs in terms of their character. The events of history are best explained by the character of the times. Mao argued that his was an age of war and revolution. Deng Xiaoping by contrast described the succeeding era as one of peace and development, which was why China could concentrate on its own growth and put off conflicts with the other major powers.

In the 2010s President Xi described his as an age 'with changes unseen in a century', meaning that this was a time of both construction and destruction, unpredictable in its patterns. This unpredictability was partly the result of the unavoidable clashes that come when one power is declining and another is rising, and he was certainly convinced that the US had decided to do what it could to choke off China's rise.

Seen through these eyes, science becomes just another weapon in a classic geopolitical conflict. China has committed to raising its own spending on science to 2.5 per cent of GDP, and now challenges the US on many fronts, prompting the other aspiring superpowers, like India, to follow suit. China, the US, Russia and India all seek prowess in artificial intelligence, genomics, smart weapons and rockets, and all are increasing military spending as fear feeds on fear and nationalism feeds off others' nationalism and, as every big nation persuades itself it is a victim, and nurtures examples of its humiliations.[10] The US has largely blocked any global rules on artificial intelligence and chips, while investing massive sums to shore up its domestic microprocessor industry. China uses both legal and illegal methods to pull in new knowledge and achieve technological superiority, from industrial espionage to investment in early stage companies, while also trying to shift the global system to its advantage.[11] Russia meanwhile has largely opted out of the global system, with its actions over bio-risks a striking sign of how the evolutionary feedback can work. Its response to the signing of the Biological Weapons Convention in the mid 1970s was to dramatically increase investment in bio-weapons, since if others were reducing theirs, a new opportunity for comparative advantage was opening up. More recently it has launched disinformation campaigns, like the claim that Ukrainian labs were in breach of the convention – all disproven by independent investigators, and manouevres to weaken the UN's inspection capacities.

So, in one plausible scenario, science becomes an ever more potent weapon in competition between nations, with blithe disregard for the risks, leading the world towards extinction, the unintended consequence of blind actions, a detachment of knowledge from wisdom. A more positive scenario draws on examples like International Partnership Against Impunity in the Use of Chemical Weapons, led by France and endorsed by forty states and the EU. In an optimistic view, well-rounded scientists are likely to move into positions of power, as adept in the ways of the political world as in knowledge. China may be the trailblazer, with figures like Chen Jining (former head of Tsinghua University, Mayor of Beijing and now party chief in Shanghai) as models. Perhaps a new generation of scientist-diplomats[12] will learn how to build on the relative successes of the Montreal Protocol and the IPCC to create aligned rules on issues such as cybersecurity or human cloning. These may even come to be linked to conditional access to the other benefits of globalization, whether new vaccines or trade. Some of the threats should make cooperation obviously sensible. As Martin Rees put it, the global village will have its village idiots, but now they have global reach, and potentially huge destructive power.

11.4 Governance deserts

For now, however, the evolutionary dynamics are contradictory. In the next century there may be a straightforward clash between the evolutionary logic for nations and that for the species. For nations it may make sense to hoard knowledge and accumulate weapons. It may seem too risky to decelerate any area of scientific advance for fear that this would confer an advantage on enemies. After all, the effective time horizons of decisions are likely to be quite short – a generation at most. In business too, short-term evolutionary feedback is likely to drive competition, even if that generates risks as an externality or spillover.[13] Both nations and businesses may, as a result, generate multiple problems that can only be handled at a higher level, even though there are inadequate capacities at that higher level to do so. The evolutionary dynamics that produced strong nation states have yet to produce comparably strong institutions of global governance. These lurch forward only after shocks and disasters, not in anticipation of them. And so, the world lacks governance of science and technology in some of the most important fields. War in space; cyber and data attacks; biosecurity; policing carbon and waste. These are governance deserts, arid borderlands without rules or predictability. In the next chapter I turn to how these governance deserts might be filled.

12

Governing global science and technology

It may seem unlikely that any kind of global governance, let alone some kind of global democracy, could hold science and technology to account or guide it. But, ten years before the UN was created, the prospects for any kind of global governance looked equally bleak. Realism meant pessimism. Yet history moves in non-linear ways, and the realists often end up being unrealistic. History curves, twists and bends, both for better and worse. So, it makes sense to be prepared, and what is impossible in one decade can become inevitable in another.[1]

One reason for cautious optimism is that there is strong global support for global institutions. Surveys consistently show large majorities backing the UN and a minority, but growing, who strongly identify as global citizens. A survey conducted for UNICEF found that 39 per cent of young people identified more with being part of the world than their own nation or region, compared with 22 per cent of the forty-plus group. With each additional year of age, people were on average about one per cent less likely to identify as a global citizen.[2]

This growing sense of global identity has been a major force in putting climate change on the global agenda and promoting humanitarian and development aid. Necessity may also force change. If any of the forecasts of the Intergovernmental Panel on Climate Change (IPCC) and others are accurate, then we face many plausible scenarios where it will be hard to survive, or at least to survive well, without much more action, and collaboration, at a global level. Otherwise, the central forecasts are for ecosystems to die off, food shortages to become normal, soils to turn to dust, and species extinction to accelerate alongside knock-on disasters in financial and social systems, all fuelling heightened competition and aggression both between nations and within them. The COVID-19 crisis was, in this sense, perhaps, a gentle rehearsal, which showed simultaneously the power of global science and how easily nations could retreat into nationalism, as each country moved decisively to guarantee vaccine supplies for its own population.[3]

The pressures of necessity may be even starker in relation to war. The prospect of war in space, with craft-destroying satellites, filling the skies with debris and damage, may be unwelcome to all of the great powers. So, too, would a worsening of cyber-attacks and cyber insecurity. The absence of any mechanisms for global governance in these fields – the last major treaty on space was signed in 1967, and there have been none on cybersecurity – is glaring. Even the better examples – like the Biological Weapons Convention, agreed in the 1970s – are seriously flawed. Since it came into force, the number of countries developing biological weapons has more than doubled, and efforts to tighten up its powers to monitor enforcement were blocked by the US in the early 2000s. It is also now seriously out of date, lacking any reference to issues such as gain of function research, genome editing and synthetic biology.[4] In short, there are glaring holes where global governance should be helping to handle the risks that individual nations cannot handle alone – from climate change to pandemics, financial crashes to organized crime.[5]

12.1 The persistence of the idea of global government

The idea of whole world governance, of sovereignty reorganized at a planetary scale (de Vitoria's suggestion of a global republic – the *res publica totius orbis)* is as obvious as it is distant. We now take for granted the deep interdependencies of our world, and have done so ever since the first photograph of Earth from space made us simultaneously see our world as a single thing and grasp its vulnerability. Even a cursory study of history reminds us that many civilizations destroyed themselves and, since the mid twentieth century, the prospect that we too might experience civilizational suicide has become a commonplace.

But what kind of institutions might avert that threat? Form, to some extent, follows function and the design of institutions is bound to reflect their most important tasks. Nations primarily concerned with defence look different to ones concerned with welfare. The same applies to global governance. Existing institutions roughly fit three types. A first type exists to enable flows. The precursors of the UN arose from the Congress of Vienna and were designed to solve very specific tasks around trade, allowing free movement along Europe's great rivers. They established principles that were later applied to postal systems and the telegraph (the International Telecommunications Union was first established in 1865, far ahead of any other global organizations). Their often prosaic work, to make it easier to transport goods or messages, set the tone for much of the everyday work of global governance – setting detailed standards for trade in wheat or steel, and later mobile phone messages, bank transfers, airline routing and safety, and such things as barcodes and html.

A second type exists to manage warfare. This was the primary purpose of the League of Nations and later of the United Nations, with the intention of prohibiting interstate warfare and, through the 'responsibility to protect' principles, wars launched by states against their own citizens.

A third type exists to perform specific tasks such as stopping famines, promoting development or helping refugees, tasks performed by bodies like the WHO, FAO, ICAO, IMO, IWO, UNHCR and many more. As we shall see, science and technology play an important role in how all of these work, from setting standards for data transfers to satellites monitoring troop movements to coordinating action on pandemics.

12.2 The right metaphors: governance as a network not a single command centre

In thinking about the possible future of global governance for science we need the right metaphors. A century ago it was assumed that a global government would look like national governments. There would be a single governing assembly; perhaps a single army; perhaps a single money; and a single permanent civil service. This is also the image of many sci-fi novels and films, which portray a council of elderly men and women speaking in portentous tones, with their own fleet of spacecraft and a supportive bureaucracy. This was one ideal of a future of science-based government: cool, calm and sleek.

But this metaphor is misleading. A more plausible metaphor imagines a network of related, cooperating and sometimes competing entities dealing with the multitude of different global tasks. This is how global governance has anyway evolved with different principles for the IMF, World Bank, UN Security Council, UN General Assembly, WHO, WTO, ILO, ITU and so on (sometimes called a multi multilateral system). Indeed, this complex picture has become steadily more, rather than less, complicated in recent decades, thanks to 'forum shopping' and turf wars, shifting patterns of legitimation, political controversies, new informal structures like the international parliamentary networks and, in the last decade, China's moves to create competing bodies like the Asia Infrastructure Investment Bank and the Belt and Road Initiative. Collaboration in some fields has been matched by aggressive competition in others, such as Donald Trump's 'China Initiative' designed to stop IP theft by China, which dampened cooperation, and the West's imposition of sanctions after the invasion of Ukraine, which helped to isolate Russia, even though research cooperation continued between Russia and China, India and South Africa. Each of these shifts has brought more trade barriers and restrictions on technology trade, making it more likely that the world will divide into competing standards and technology systems.

But, even within such a messy context, there is still space for institutions that can solve common problems where interests continue to be aligned – from migration to epidemics, drugs and organized crime to cybersecurity and security.[6] Already, air safety, IP protection and many other topics have specialized organizations that have arguably proven more adaptable than the bigger, more politicized organizations. The same is true of global financial regulation and tax alignment and action to reduce the role of offshore havens and evasions.

The most interesting shift is the rise of bodies dedicated to orchestrating knowledge to help the world think and act and using scientific knowledge as a lingua franca. These are the everyday expressions of science in governance, and a first attempt at a kind of global metacognition. The IPCC became the most visible expression of this ethos, set up by the UN (the World Meteorological Organization and UNEP) in 1988 to connect scientists and make forecasts on climate trends and their implications. It has three working groups and a taskforce; each group has a chair from a developed country and a chair from a developing country and provides assessments of the state of the world every seven years or so. These aim to monitor and make sense of climate-change impacts, adaptation, vulnerability and mitigation. The most recent synthesized over 14,000 peer-reviewed studies around the world, with over 70,000 comments from experts and drafts agreed by scientists from 66 countries, and then reviewed by representatives of 195 member governments.

Many of the more recently created bodies – like the Intergovernmental Science-Policy Platform on Biodiversity and Ecosystem Services (IPBES) – have also prioritized generating and sharing knowledge in order to influence decisions. A typical example is IUCN, the International Union for Conservation of Nature. It has 1,400 institutional members that include nation states and NGOs, scientific and business organizations, and provides analysis and ideas (some of which end up as conventions). Gavi, the vaccines alliance, is another example: created by the Gates Foundation, it includes national governments and UN agencies on its board, though in a minority, and its main task is to orchestrate knowledge. The Climate and Clean Air Coalition brings together hundreds of governments, NGOs, businesses and cities, again without any formal grounding in a treaty. These are part of a lattice of collaborations that include CERN, Europe's particle-physics laboratory, and the James Webb Space Telescope, as well as the many COPs, such as COP15 on biodiversity, which agreed in late 2022 a blueprint to halt and ultimately reverse the decline of species and the degradation of ecosystems.

Many types of global commons now sit alongside ones like the IPCC. There are vast shared databases, like the Protein Data Bank created in 1971, which in 2020 made possible Deep Mind's leap in machine learning to discover new proteins. Something of the spirit of the scientific ideal of a global commons committed to a truly common good enabled the rise of research collaborations and projects like the HGP and the International Atomic Energy Agency (IAEA), the dominant institution in international radiation dosimetry. The 1972 United Nations Conference on the Human Environment, organized in Stockholm, is widely seen as a momentous event in environmental history, and a model for other fields.

Many of these initiatives emphasize observation: global tools to observe the world in real time, increasingly trying to combine commercial data with public data (for example, from mining, agriculture and telecoms companies). IPCC is the most prominent; others include the Sloan Digital Sky Survey, trying to map the entire universe, and Metasub, mapping pathogens on urban metro systems. Others organize evidence, like MEDLINE, the US National Library of Medicine's database containing nearly thirty million references to journal articles in forty languages in life sciences, with a concentration on biomedicine.

New data tools help to map and visualise these complex new patterns of research, cooperation and impacts.[7]

Yet, in some fields, there are not even the most basic observatories of this kind. Stronger registries and inspection agencies are badly needed to monitor hazardous biological research, AI or weapons, and these are the precondition for effective global governance. The field of artificial intelligence, for example, lacks any institutions remotely comparable to the IPCC. No observatories can draw on the inputs of thousands of scientists to monitor, assess and explore scenarios. Instead, as indicated earlier, there are dozens of small initiatives[8] and the occasional flurry of open letters and hand-wringing.

Even where there are strong institutions devoted to observing and assessing, these lack powers of enforcement: the stage of action that, as I showed in Chapter 1, is a vital part of science. One future of governance would build up the formal authority of such bodies, giving them more 'state-like' powers to deploy large sums of money or to enforce international law; for example, bans on the misuse of gene-editing or potential bioweapons. For combatting climate change, many have argued for an agency with powers not just to monitor nation states' carbon reduction plans but also to introduce sanctions if they fall short – through less favourable trading terms or access to credit. The Paris Agreement succeeded in achieving a consensus because it avoided legally binding commitments (its predecessor, the Kyoto treaty, had many flaws but at least had the status of law), but that may prove insufficient as global warming intensifies.

We may need a Global Data Agency to gather and manage data on the global goals, ensuring common standards and quality, and with some powers to mandate contributions of essential data from governments and businesses. We are likely to need a Global Agency for Machine Intelligence, able to monitor and inspect powerful artificial intelligence, to enforce moratoriums or penalize risky actions, generalizing some of the legal frameworks arising in the European Union and China. In the future these might even operate on a 'national credit' system, with nations gaining credits for succeeding in cutting carbon or abiding by international laws. These credits would give them preferential access to trade and finance, matched by commensurate penalties for those who act as bad global citizens.[9]

Any such institutions are hard to imagine in the climate of the mid 2020s. They would need the backing of major powers, which for now would rather hoard their power than share it. But there are many scenarios in which new institutions could serve the interests of many countries, including the major powers, and could steadily rise up the hierarchy of tools that global institutions can use: starting with open observation of the facts of the situation, moving onto formal powers to inspect facilities, algorithms and data, registries, and methods of assurance to ensure safe processes are being followed and, finally, penalties for infractions, whether imposed on individual institutions or whole nations.

The European Union is a partial model and looked equally implausible twenty years before it was founded. The diplomat Robert Cooper captured its essential idea in the concept of a 'post-modern state' that achieves security through transparency rather than secrecy, and in which states acknowledge their interdependence by pooling power and money and abiding by mechanisms for mutual coordination. This model, which can also bring with it calibrated penalties and incentives, represents a fundamental shift away from the classic state built on secrecy, hard borders and a claim to unlimited sovereignty.

Yet the post-modern state has to coexist with classic modern states interested in force, sovereignty and military power (which continues to be the large majority of sovereign nations), as well as with what Cooper called 'pre-modern states', areas that lack the functioning capacity of government and have, in some cases, collapsed into civil war or chaos.[10]

12.3 Science and the Sustainable Development Goals

An irony of the many existing and possible bodies described above is that there is no governance of science itself – no place to debate whether the world's R&D is adequately directed to the tasks the world believes to be most important. Instead, there are glaring skews: the continuing priority given to military research, given additional impetus by the wars and conflicts of the 2020s; within fields like health the priority given to the less important diseases of the rich world, and within agriculture the skew to particular forms of agribusiness, and so on.

To counter these distortions, the world needs a more synthetic, hybrid approach to science, politicizing it in the best sense of the word, and providing a bridge between the world of facts and the world of determining what matters and what is to be done. Again, a starting point is to bring the facts into the open, with a global observatory for science and technology, responsible for gathering and harmonizing data, making forecasts, and attempting to overcome the secrecy that surrounds R&D for military and intelligence purposes.[11] Such an observatory could analyse how R&D relates to global disease burdens; the development of R&D capabilities in lower-income countries; the potential negative impacts of and inequalities generated by technologies; different innovation pathways; and how far individual nations have a good alignment between their R&D and the SDGs.[12]

An observatory of this kind would make it easier to bring together partnerships and assemblies of key players in specific fields, gathering around priorities such as energy, child malnutrition or water, and generating shared maps of funding allocations to avoid duplication or tackle gaps. These 'constellations' would typically bring together national bodies, major development funders, civil society and science. There are many kinds of partnership already, for example around malaria, access to water or gender equity. Usually their tasks are time-limited rather than permanent – for example, addressing intense phases of a problem such as conflict reconstruction, drought or famine, a refugee surge or a financial crisis. Most combine private funding (primarily philanthropic) and public money.

Other partnerships need to be more permanent. Disability, for example – an issue that affects more than a billion people worldwide – is a prime candidate for a new constellation to coordinate research, development and commercialization: funding science and technology to address needs like sight, hearing, mobility; running testbeds and labs; promoting policies and new rights (including in the labour market); advocating for voice and expression. Food is also a good example because of the range of existing bodies such as the Commission on Sustainable Agriculture Intensification, processes such as the International Assessment of Agricultural Knowledge, Science and Technology for Development, and gatherings such as the UN Food Summit. A more formal constellation could open up debate about alternative pathways, including the merits of precision agriculture, GM seeds and insect growth regulators on the one hand, and more agro-ecological methods, such as rainwater harvesting adapted to local conditions, on the other.[13]

A next step up from constellations is formal pooling of budgets, which is what organizations like the World Bank and UNDP already do. CGIAR (originally the Consultative Group for International Agricultural Research), for example, has operated a pooled budget since the 1960s, amounting to over half a billion dollars each year, and linking foundations including Rockefeller and Ford with major public donors. After playing an important role in the 'Green revolution' of the 1960s, much of its work focused on the genetic development of crops, which sparked controversy. Other examples include the Global Fund (which has mobilized around $4bn each year to support projects dealing with AIDs, TB and Malaria, and spent nearly $50bn since 2002)[14] and the Global Innovation Fund, a recent collaboration between the UK, Sweden and US governments, foundations such as Omidyar and companies such as Unilever.[15]

These suggestions emphasize action. But talk also has its place. It is easy to dismiss conventions and summits as empty talking shops. But such events play a crucial role in creating communities of shared purpose and understanding, as well as in catalysing or provoking wider social deliberation over the steering of policy. This is true of the COP series, G7 and G20 and others, which – for all their imperfections – contribute to an alignment of purpose. The failure to align R&D with the SDGs partly reflects the absence of such discussions. The OECD has its Global Science Forum[16] and UNESCO has its Global Observatory of Science, Technology and Innovation Policy Instruments[17] but neither feed into aligned decision-making. The same is true of gatherings like the STS Forum[18] and, more recently, of GESDA focused on anticipating future science trends.[19] Talk and conversation can seem like indulgences. But humans are social animals and we have found no other way to create new senses of shared purpose.

12.4 A new economic base for global bodies: taxing global public goods

One of the biggest challenges that all global bodies face is funding. None have the power that nation states depend on – the power to raise taxes that derives from their monopoly of coercion. Instead, global organizations have to pull together funds from governments and then become dependent on the bigger donors. Increasingly, super-rich philanthropists have filled part of the space – which is good in terms of addressing needs, but unhealthy in representing a return to pre-democratic models of power without accountability.

An alternative would more deliberately sort out the economic base of global governance, using global public goods to fund global public goods and, more specifically, using global public goods that are the result of science to fund future science. Specifically, this would mean raising taxes or licence fees for such things as geostationary orbits, electro-magnetic spectrum, access to natural capital, oceans and the like, and potentially air traffic routes and landing slots, and the major seabed communication links, and using these resource flows to fund global action.

Getting this right would greatly transform the psychology and confidence of global institutions (though more funding would also need to be matched by transparency and strict auditing to ensure efficiency). Making this happen would depend on leadership from the major powers which, if short-sighted, would see it as a threat. But the benefit would be more effective global action, from which we would all gain in the long run.

12.5 Global democracy and legitimation

Many bodies now exist that could provide the kernels for global governance of science and technology. But how would they connect to politics or the public of the world? In one image of the future, all of the world's citizens would vote for a global president, or a global assembly, which in turn might set priorities for science and technology. This is almost certainly neither practical nor desirable, and would lead to domination by the most famous, not the best suited to lead, or by the most populous nations, which would dominate the ones with smaller populations. Awareness of similar risks led large federal states to create hybrid models, like the US Senate, which gives equal representation to each state to balance the Congress that is weighted to population.

The most important issue both for the system as a whole and for its parts is legitimation: without legitimacy they cannot act, raise money or expect compliance. But their legitimation tasks are different, with relatively little spill-over of legitimacy from one organization to others. The legitimation needs of highly technical standards bodies are very different from ones involved in peace-keeping or trade. So, we should imagine not one person one vote but rather hybrids that give some role for the peoples and some for the nations, but with a variable geometry to reflect the range of tasks (and in some cases the reality of military and financial power), and to reflect what in pre-modern times were called the estates, which, in this case, would include business, civil society and science. These would better reflect the reality of the current world system than the assumption that national governments alone can represent the diversity of views and interests of the world.

The biggest roles for reformed democracy in the global system, and in relation to science and technology, should come in the early and late stages of the democratic cycle: the stages of proposing and nominating issues, suggesting ideas and scrutinizing options that come before the moment of decision. Democracy has a lesser role to play in the stages closest to decision, which is bound to involve a harder-edged assertion of interests and more secrecy. But it again becomes relevant in the later stages of monitoring and learning. These are all very amenable to the use of digital technologies, as the UN discovered when over 10m people took part in its first global consultation on SDGs. There could also be a role for sortition, with random samples of the world's population providing inputs to strategic discussions. It would not be hard to imagine a global version of the UN assembly using analytics and visualizations to map citizen inputs on emerging issues, commenting on ideas and scrutinizing the reality of actions.

The decisions themselves could still be taken by much smaller groups in the very varied forms of governance likely to be needed for different fields. But changing the environment for those decisions would have a big impact. Indeed, this is the great lesson of democracy in nations. Periodic votes matter but, just as important is a dense hinterland of facts, ideas and argument, followed by transparent scrutiny. The ecosystem is as important as the formal procedures.[20]

12.6 A renewed United Nations founded on knowledge

When the UN was created, Article 109 committed it to regularly reviewing its operations and charter every decade, a wonderful principle of self-reinvention, but one that has not been invoked since the 1960s.[21]

A useful thought experiment is to take Article 109 seriously and imagine that the United Nations was being invented in the 2020s rather than the 1940s. Then the priorities included stopping interstate war, reshaping flows of finance and helping refugees. A United Nations being built now would place data, science and knowledge on as prominent a footing as finance, reflecting an economy in which the most highly capitalized companies are largely based on data and knowledge, rather than finance or oil.

So, we would not just have a World Bank and an IMF but, as indicated earlier, a clutch of institutions to mobilize knowledge of all kinds and make the most of the world's brainpower, all aimed at accelerating the achievement of the SDGs. We shouldn't be too cautious. Big changes in governance always look impossible and unlikely – until they happen. But once they've happened they appear obvious and inevitable.

PART VI

The Problems of Meaning
Synthesis, Wisdom and Judgement

13

Science, synthesis and metacognition

Bertrand Russell was one of many who believed that all meaningful questions could be solved by science: that, as we are part of nature, our scientific method should be relevant to any kind of problem. The rest he thought was essentially fuzzy and meaningless. The job of philosophers was to clarify concepts, not to get lost in vague metaphysics.

But these attempts to cull other ways of thinking have always fallen short. Within limits they are entirely desirable. Taken too far they exemplify the gap between cleverness and wisdom. The world around is vastly more complex than our brains or theories can grasp. So, we need to remain humble about our knowledge and to protect ourselves from overconfidence – testing, experimenting, doubting, fighting against our own tendencies to seek confirmation.

Humility is also needed because of the gap between science and meaning. The scientific method can drift into nihilism. Doubt, scepticism, showing the emptiness of others' claims, are marvellous methods for driving us to greater clarity. But they then make it harder to fill up the space that is left, the space in which we live our finite lives.[1]

There are now few examples of philosophical questions that do not have some empirical, and scientific dimension – how we imagine, how we feel, how we judge, the bases of right and wrong; how we imagine our relationship to the biosphere, or future generations. Yet science alone can answer none of these questions. Humans are designed to seek meaning. Science denies it, indeed it shows the cosmic irrelevance of humans whose meaning comes from the existential facts of life and death, and, while we are constrained by being in a body in a place and time, knowledge is oblivious, rootless, disembodied.

Science can never be a complete world view, only a part of a world view. It can never be a universal philosophy, as Bertrand Russell hoped, but only a partial one. On its own it can never provide enough guidance to shape decisions, and to synthesize. And a political system solely governed by science would have a hole at its heart. It could explain and guide. But it would struggle to create compelling meanings – answers to questions about why, about belonging, love, fear and hope.

Ibsen's classic play *Enemy of the People* was an early exploration of the dynamics of science and politics. Its lead character is Dr Thomas Stockmann, the medical officer of a recently opened spa. Tests confirm suspicions that the spa water is contaminated with bacteria, which, it turns out, originate in his father-in-law's tanning works. The results threaten the spa and the town's economy. Dr Stockmann chooses to speak the truth. But many powerful interests stand against him, from the anti-government newspaper (which decides not to print his article on the tests, on the grounds it would do more harm than good) to the mayor. In response Stockmann becomes ever more a crusader and, at a crucial town meeting, he denounces the 'colossal stupidity of the authorities' and the small-mindedness of 'the compact liberal majority' of the people. Science, he asserts, is simply superior and has to be believed. Insulted, the public turn against him and denounce him as an enemy of the people. The play ends with him asserting his strength as the man who can stand alone in defence of truth.

The play has become a staple of the repertoire. It works as drama. But its performance in Beijing in September 2018 showed that it still had power as more than entertainment. The audience showed passionate support for Stockmann and shouted aloud critical comments on the Chinese government, calling for personal freedom. Government censors decided to block the tour unless the script was adjusted.[2]

So here we see a scientist and politics at odds with each other. But is Stockmann such a hero? Ibsen was ambivalent about his main character: he recognized his blind spots and, of course, that made him more interesting as a figure. He seems unable to listen to others' views; he takes no account of other values and interests. He seems to assume, like many, that scientists are innately virtuous.

The play also calls attention to the question of synthesis. The audience in Beijing was right to want the truth to be spoken. But that is the beginning of wisdom not the end. Because in the case of Ibsen's play, as in so much of life, there are parallel truths, not single ones. Whether or not the water is contaminated is in this sense straightforward. Either it is or it isn't, and it's useful to be able to distinguish truth from falsehood in this kind of case. But what next? How to take account of the economic interests, the jobs, the survival of the town? These types of fuzzy, conflicting judgements are ones we all have to make all the time. Few choices are easy to deduce from a few principles.

Scientists often assume that the knowledge they feed into the governments is uniquely authoritative. It's grounded in observation, theory and experiment. A scientific fact trumps any other kind of fact. A more scientific government is better than a less scientific one, and more science, more advice, and more education are all for the good. This is a position for which I have much sympathy, certainly when pitted against its opposites.

But it is also radically wrong – an epistemological error that derives from a mistaken understanding of the relationship between knowledge and action. At its simplest, the point is that a single piece of new knowledge does not entail new action. Instead, the task of understanding may require many different kinds of knowledge, even if the resulting action is simple.

Imagine you are talking to a friend going through a personal crisis, and you wish to give advice. Your mind will explore multiple options – trying to grasp what's really happening, its causes, its severity, its possible consequences – before deciding what to say. The same may be true of a decision about pensions, or buying a house. Knowledge goes wide before narrowing down for action.

This becomes apparent if we look in any detail at the role of science in government. A good example is climate change. By the mid 1990s there was a fairly wide consensus about the realities of climate change, its human causes and the need to avert future disaster. In the early 2000s I was running the UK government Strategy Unit and part of our task was to turn this consensus into policies, both specific policies for topics such as renewable energy and waste, and also more generic approaches, such as making the case for 'resource productivity' to sit alongside more traditional ideas of labour productivity in economic policy-making, so as to encourage attention to flows of resources and materials.

We were helped by a wide network of scientists, engineers, committees in parliament, advocacy organizations – many pressing government to do more and faster. These were rich in expertise. The result was an evolving cluster of policies. Some focused on energy – shifting faster to renewables and away from coal in particular. Some focused on transport – pushing towards more energy efficient cars and trucks. Some focused on buildings, with strict standards for new buildings and generous subsidies for retrofitting old homes.

Not all were successful. But taken together they achieved a fairly remarkable result. Production-based CO_2 emissions went down nearly 50 per cent, while consumption based emissions (which takes account of the origins of goods and services consumed) went down close to 40 per cent over the next twenty years. It's easy to argue that these were not enough (overall the world economy is decarbonizing at a rate of roughly 1.5 per cent a year, far less than is needed to avert climate change). But the results are good examples of science as a prompt for necessary action.

However, closer examination shows that although science was the prompt, the successes, such as they were, relied on multiple kinds of knowledge being mobilized in parallel. They relied on engineering, economics, psychology, detailed tacit knowledge within particular industries, to name just a few, usually with much more 'techne' than 'episteme'. The science on its own could generate only a small fraction of the knowledge needed for action.[3]

A similar pattern could be seen in the COVID-19 pandemic. Science helped governments to see the risks of the pandemic and modelling helped them to grasp just how quickly the disease could spread and overrun hospitals. But the scientists and scientific advisers could only offer a limited set of insights in order to help governments act. Seen through the eyes of the decision-makers all of the following kinds of knowledge were relevant:

- **Statistical** knowledge (for example of sample-based infection levels or unemployment rises in the crisis).
- **Policy** knowledge (for example, on what works in stimulus packages).
- **Scientific** knowledge (for example, of antibody testing).
- **Disciplinary** knowledge (for example, from sociology or psychology on patterns of community cohesion).
- **Professional** knowledge (for example, on treatment options).
- **Public opinion** (for example, quantitative poll data and qualitative data).
- **Practitioner views** and insights (for example, police experience in handling breaches of the new rules).

- **Political** knowledge (for example, on when parliament might revolt).
- **Legal** knowledge (for example, on what actions might be subject to judicial review or breach Human Rights Conventions).
- **Implementation** knowledge (for example, understanding the capabilities of different parts of government to perform different tasks).
- **Economic** knowledge (for example, on which sectors were likely to contract most).
- **'Classic' intelligence** (for example, on how global organized crime might exploit the crisis).
- **Ethical** knowledge about what's right (for example, on vaccinating children who may have relatively little risk from a disease).
- **Technical and engineering** knowledge (for example, on how to design an effective tracing system).
- **Futures** knowledge (foresight, simulations and scenarios, for example, about the recovery of city centres).
- Knowledge of **lived experience** (the actual experience of patients, families, front-line staff and others).

There are many other ways of constructing or structuring such a list. Each type of knowledge tends to have its own professions and institutions, jargons and languages, and most struggle to understand the others. But, however such a list is constructed, two crucial points follow from any recognition of the diversity of types of knowledge that are relevant to decision-making in governments, or to put it another way, to the needs of political power.

The first is that there is no obvious hierarchy or meta-theory to show why some types of knowledge might matter more than others. Leading figures in particular fields may feel that it is obvious why 'their' knowledge is superior to other types of knowledge and a theory of physics that can be shown to fit data in a very wide range of contexts is indeed superior as theory to knowledge that is more context specific. But, seen from the perspective of action, this superiority disappears. What matters is what works, or what is useful. Seen through this lens it is impossible to prove the superiority of any one kind of knowledge convincingly. It may be superior for a particular task at a particular time, but it is very unlikely to be superior in any more general sense.

To judge which kind of knowledge to apply to which task requires skills in metacognition, intelligence about intelligence. Here for example is a comment from Sweden's head of public health, which contrasts with the narrower messages which came from science advisers in other countries:

> We have actively resisted school closures to whatever extent possible and this is because schools are extremely important from a broad public health perspective. Schools are perhaps the most important institution that society has when it comes to creating good public health. And, if you close them down, you'll get very many negative effects. And you can see this more and more, in the media and in reports from around the world, that what is being created now is a lost generation of children who haven't been able to go through school like you'd normally do.

His judgement was rooted in evidence and logic; but it was radically different from judgements made in many other countries, and it attempted to see the pandemic not just through the lens of infections and mortality (and, in retrospect, Sweden's policy appears to have been one of the most successful).

The status and influence of these different sources of knowledge may correspond only loosely to what is needed at any particular time (a point confirmed in a piece by past and present scientific advisers).[4] No formal models or heuristics exist to show which kinds of knowledge, and which models or frameworks, are relevant for which tasks and when, though there are many methods for summarizing or linking different kinds of knowledge, using knowledge graphs and other tools. Instead, the metacognitive ability to know what to apply and when depends on a kind of wisdom that can only be gained through some familiarity with these different kinds of knowledge and their application, and through experience.

How, then, should the often-contradictory signals coming from the different kinds of knowledge listed above be synthesized? It's sometimes assumed that this job will be done by politicians, helped by political advisers. But they rarely have the time or skills to do this well. Sometimes it's assumed it will be done by senior civil servants – but again they may or may not have the skills to do this well (and are likely to be more confident dealing with issues of law and economics than with data or science). Even the most sophisticated accounts of science advice and knowledge brokerage still present it as an input and support for decisions that are taken by others, leaving the crucial moments of decision as a kind of black box.[5]

The result is often an excess of advice and a deficit of synthetic capability. There may be marvellous inputs from brilliant scientists, but only very limited capacity to use it. The key point is that no system can function without some capacity to synthesize multiple kinds of knowledge.

13.1 Types of synthesis

Synthesis never happens in a straight line – it is never purely deductive or logical and, indeed, it is never stable because if you change the weightings you change the results. So it is always contingent, situational and contextual and there are many types of synthesis. It is possible to synthesize downwards – putting multiple things into a single metric (like using money as a standard measurement, which is what cost-benefit analysis tries to do, or the use of QALYs as a single metric for judging health interventions). It's possible to synthesize upwards with a frame or theory that makes sense of multiple and otherwise confusing things. In 1913 the physicist Jean Baptiste Perrin wrote of breakthroughs that 'explain the complex visible, by some simple invisible'. Einstein's theory of relativity and Darwin's theory of evolution are both examples, which made sense of many confusing current theories and observations. You can try to synthesize forwards – drawing on many inputs to decide on a course of action or strategy, such as a military campaign or a plan for mass vaccination (as Richard Rumelt put it in his book on strategy, 'reducing the complexity and ambiguity in the situation, by exploiting the leverage inherent in concentrating efforts on a pivotal or decisive aspect of the situation'; or to synthesize backwards, making sense of historical patterns in a narrative that distinguishes the critical factors (we understand our lives backwards but are condemned to live them forwards). It's possible to synthesize with analogy, for example, seeing the Earth as a single organism (Gaia), seeing a pandemic as like a war, or seeing the spread of an idea as like a virus. In the everyday work of governments other options include synthesizing through a heuristic with simple decision rules that can work most of the time (e.g. a monetary policy target, or an injunction to business start-ups to focus on cash).

How, then, can syntheses of this kind be generated? In food, music, poetry and other fields there are an infinite range of methods. But, for understanding or acting on complex systems, and the kinds of issue that accumulate on the boundaries of politics and science, there are fewer. Instead, any such work of synthesis is likely to need the following stages:

1 **Mapping** relevant factors, causal patterns and relationships and attempting to put them into a common language or common models.
2 **Ranking** these inputs, models or insights in terms of their explanatory or predictive power.
3 Attempting **mergers** or combinations (sub-syntheses).
4 Clarifying **trade-offs** and **complementarities.**
5 Clarifying **knowledge and power**, i.e. which causal links are well or badly understood, and which ones are amenable to power and influence.[6]
6 **Jumping to new concepts, frames, models or insights**, that use these inputs but transcend them.
7 Finally, **interrogating and assessing** these new options and judging how much they create or destroy value.[7]

This is often a circling rather than a linear process. It involves trying out, exploring and interrogating on the way to a viable answer. It's not a one-off exercise of the kind offered by multi-criteria analysis or cost-benefit analysis.[8] It may start with one set of models, discover these are inadequate, and then bring in others. Often diagnosis is best separated from prescription to avoid the risk of participants becoming too attached to potential options.

In some fields models can play a useful role, whether to connect the work of analysts and policy-makers[9] or to combine inputs from many discussions.[10] So too can structured meetings, and there are many methods available for organizing meetings to do synthesis[11] or to make sense of complex systems and their options for transition.[12]

Values and ethics run through every stage (and can't simply be added in as another ingredient in 1 and 2). They shape what questions to ask at the beginning and the choices offered at the end. The sixth stage invariably requires different methods – more lateral and visual – since the human brain cannot grasp systems through linear prose or equations alone. It often also involves alighting on a narrative, a framing story that makes sense of complex patterns.

In short, the point of these processes is to clarify how far different goals or pathways can be aligned or integrated in novel ways; when there are unavoidable trade-offs; and when outcomes are incommensurable or conflicting.

The work of NICE is a good example of evidence synthesis focused on action: it guides the UK's National Health Service commissioning decisions helped by a tool for synthesizing downwards – assessing all treatments in terms of their likely impact on QALYs.[13] The IPCC is a good example of institutionalized synthesis for understanding, at least of the first four stages (it has less authority to do the other three). It produces formal syntheses (most recently AR6); it makes use of multiple models and processes of deliberation, seeking a collective intelligence that is superior to any of its component parts, as well as preparing transformation and mitigation pathways (though one its problems – which mirrors governments' use of tools such as cost-benefit analysis – is that these often become overly rigid and fixed, and fail to adapt to shifts in the environment). But the IPCC doesn't cover synthesis for action – i.e. orchestrating global knowledge on what works in relation to decarbonization. Judicial inquiries are another method for synthesis (mainly synthesizing backwards, i.e. making sense of an event, as well as making recommendations for the future).[14]

The good examples take care not to see patterns too quickly or jump to conclusions (a very common problem in every field, from medicine to policing). They also combine models and modes of thought – not just linear logic and models but also visualizations and simulations since we can often see complexity more easily than we can grasp it logically.[15]

Not everything needs synthesis. During a pandemic, for example, many actions can be taken without having to take much account of their broader implications – including testing and vaccination. The best response to a complex problem – even if it has a single cause, like a pandemic – may be an assembly of multiple elements rather than a single approach. To over-synthesize them would be inefficient. Instead, the key is to see where there are important linkages and interrelationships, or where decisions have much broader impacts (such as lockdowns) and focus on these: enough synthesis, but no more.

This is a familiar point in engineering and technology, which involves assemblies of multiple sub-systems, for example, to make a car, a computer, a building or an aeroplane. The interfaces matter of course. But there is no need for every component to follow a single logic – and attempts to reinvent everything from scratch in an integrated way are usually sub-optimal.

Nor should everyone try to be a synthesizer. We need deep knowledge, rooted in disciplines, which comes from people devoting their lives to the details. Indeed, most people should prioritize depth over breadth.

But we need many who are synthesizers, adept at metacognition, and we need everyone to be aware of the limits of their knowledge, where their domain ends and others begin. We also need the distinctive mindsets suitable for synthesis, and for collaboration with others from very different backgrounds and interests. The greatest enemies of collaboration are distrust, arrogance and hubris. Without trust, people will tend to hide and hoard information. Those who believe that they are uniquely endowed with insights or abilities will tend to be poor collaborators and poor listeners. In highly collaborative environments, by contrast, people leave their egos and narrow interests at the door and commit to a larger common interest. They recognize the limitations of their own knowledge and perspective. This is where ethics, personality and style intersect, and where synthesis has a moral dimension beyond its technical components.

13.2 Science and wisdom

Immanuel Kant wrote that 'science is organized knowledge, wisdom is organized life'. Wisdom is what we most want from political leaders. We assume that they can access technical knowledge – but it is knowledge about that knowledge, the ability to reason ethically and judge what perspectives are right for which situation, that is crucial.

This has been a long-standing challenge for science. One example of many was the Nuremberg Code that was part of the verdict in the Doctors Trial in 1949 which led to seven of the Nazi's leading doctors being hanged. The trial had looked at the many ways in which medicine was caught up in brutal experiments. It aimed to assert ethical principles to guide experiment, including principles of consent and purpose. And, although the Nazis were the prompt, the Code was also a response to the US's own ethical laxness, particularly in experimenting on the mentally ill.

A parallel example was Japan, where the military physician General Ishii Shiro had created a secret biological warfare programme based in Manchuria in the 1930s, which ran experiments with infectious disease on live humans, most of them Han Chinese. However, in this case, because of Cold War paranoia that the USSR might be ahead, the Japanese scientists were secretly given immunity from war-crimes prosecution by the US and UK in return for information on their biological experiments, despite evidence that 200,000 Chinese civilians died from germ attacks.

So what might wisdom look like in relation to science three-quarters of a century later? To help answer this question there is now an emerging science of wisdom that attempts to draw together evidence from psychology or from many cultures.[16] It shows that wisdom tends to be associated with particular behavioural traits: calm, detachment, avoidance of impulse and an ability to see multiple perspectives.[17] It includes depth of knowledge – familiarity with bodies of knowledge, codes, symbols and disciplines, and including tacit as well as explicit knowledge. This knowledge is a combination of models (theories that state 'if this, then that . . .') and facts. Ignorant wisdom is a contradiction in terms, which is why science forms part of wisdom. But it doesn't follow – as Plato, Hayek and others assumed – that more knowledge automatically leads to more wisdom.

Indeed, wisdom also entails recognizing what's missing, the crucial data that may lend a very different perspective, and it involves knowing the limits of knowledge: that we can never fully get inside an object, another person, an historical event, or the meaning of a work of literature[18] and that we can easily become trapped by the categories we use.

Recognizing how hard it is to fully grasp how things are, 'the feel of not to feel it' as Keats put it, involves a degree of humility that is at odds with the blithe assumption that everything can be analysed. Einstein wrote about this well in 1916: 'Concepts that have proven useful in ordering things easily achieve such authority over us that we forget their earthly origins and accept them as unalterable givens . . . the path of scientific advance is often made impassable for a long time through such errors.'

Evidence on predictions shows that people with deep knowledge can be poor at predicting what will happen: less knowledge but more openness outperforms deep expertise. Sometimes research advances fast without much knowledge of causation, of why things work rather than whether they work.[19] We know that confirmation bias can be even more marked among the highly educated, who become skilled at incorporating any new piece of information into their world view, rather than using it to challenge and update their world view. Through habit and repetition our minds become accustomed to particular ways of thinking, and the longer one has been an expert in a field the harder it may be to accept new ideas and frameworks.

These insights give us what has become a common approach to wisdom in much of academia (other than psychology), summarized in the widely used DIKW framework, where data, information, knowledge and wisdom are seen as a hierarchy. In Scott Page's work, for example, the essence of wisdom is the ability to apply multiple models to understanding situations or problems, and then to choose the most appropriate models to guide decision and action.[20] 'Wisdom requires many model thinking . . . when taking actions, wise people apply multiple models like a doctor's set of diagnostic tests . . . [and] construct dialogue across models, exploring their overlaps and differences'.[21]

This gives us a framing for wisdom very similar to the ancient world. Aristotle distinguished the elements of knowledge I used earlier to describe the relationships of states to science: *episteme*, the logical thinking that applies rules, *techne,* the practical knowledge of things, and *phronesis*, which is practical wisdom (sitting alongside *sophia*, its more theoretical and abstract counterpart), and suggested that each has its own logic of verification. Episteme can be verified through logic or formal experiments. It only takes one counter-example to disprove a rule or hypothesis. Techne is tested by practice: does something work or not? Phronesis, on the other hand, is determined by context, and can only be verified through applying it to choices and learning step by step whether decisions really do turn out to be wise or not. It usually includes an ability to reason ethically and apply ethical principles to new situations, making judgements about right and wrong.

Some of these judgements are cognitive – and are essentially about knowledge and reasoning. But, crucially, others are non-cognitive, involving emotion, empathy, compassion and intuition, and the stance taken with respect to the people or the situation. Ethics in other words involves both justice *and* mercy, reason *and* feeling, detachment *and* commitment. Indeed, this is one of the reasons why in many traditions it is thought that experience of suffering and setbacks can enhance wisdom, transforming it from something that is only cognitive into something richer.[22]

Ethics can be thought of as just another kind of knowledge, but this gets it wrong. It is not an accumulation of facts, in the way that physics and biology are. It is a process, a verb not a noun. This is why attempts to codify ethics become either bland or misleading (like the many hundreds of lists of ethical principles produced around artificial intelligence, few of which had any impact in the world). And, of course, science and ethics can clash. Would the world be a better place if some discoveries had not been made – from Zyklon gas to hydrogen bombs? Quite possibly.

An opposite, and even harder example, is torture. A common view is that there is no evidence that torture works, even though in fiction and films it often does. This is a comforting position. It implies that 'is', the facts of the world, and 'ought', the ethical perspective, are aligned. But an alternative, and controversial, view suggests that there is some evidence that torture can be effective.[23] I'm not an expert on the evidence on torture and would much rather live in a nation that proscribed it. But I am struck that most of the writing on it seems to select evidence to reinforce a previous position, whereas wisdom surely lies in being able to recognize that 'is' and 'ought' will often clash.

Another crucial element that links into the role of ethics, and the looped nature of wisdom, is sensitivity to the long view. This is the ability to grasp the relationships of the present to both past and future, to see issues in their temporal context, and to spot what future potential lies in present things, whether seeds, landscapes, people or societies. Perhaps wisdom has to involve some sense of what today, and the dilemmas of the today, might look like from the perspective of the future.

It should already be clear that the science of wisdom points to a much broader perspective than any simple equation between more knowledge and greater wisdom. The US NIAID Director Anthony Fauci proclaimed that 'science is truth' – an understandable reaction against widespread lies and conspiracy theories. But it would be wiser to say that science is a search for truths and is never the whole truth.

Indeed, this takes us to the essence of wisdom and the essence of a healthier relationship between politics and science. One way of thinking about wisdom is as perpetual learning – a willingness to test and improve models for understanding the world. This is the spirit of the Royal Society's slogan *nullius in verba* and points to continuous loops of questioning and learning. These loops parallel the Bayesian inference that underpins much artificial intelligence and data science: first you decide on a 'prior' or estimated fact, along with an estimated probability; then you observe the true facts; then you adjust your model, and your probabilities, accordingly.[24]

Yet when it comes to the interfaces between science and politics this is often missing. The traditional science advice view argues for gathering a diversity of types of expertise, ideally combining natural and social sciences; having robust processes for argument and interrogation; transparent and competing models; and then feeding advice into processes – often run by politicians – that are then thought of as entirely separate. The politicians are then held accountable for their decisions. But it's rare for the experts to have to make explicit predictions and then to be held to account for their accuracy. This is perhaps one reason for the widespread view that the experts are too powerful; that the solidity of their knowledge is often exaggerated; that they smuggle values into their advice; and that they need to be demoted from their pedestals.

A better approach is to make these loops of reasoning transparent and shared, with experts making explicit predictions about 'if x, then y'; decision-makers doing the same; and explicit processes for learning when x leads to z rather than y. This implies much more visible and open processes for advice; deliberate use of multiple disciplines and frames, as well as multiple models; recognition of uncertainty; and rapid, and also visible, taking stock of what did or didn't work.

This will be uncomfortable for experts (and for other university professors like me). We will be more exposed and more accountable. There are surprisingly few moments when experts are questioned in public about what they got wrong or right, like the one when Queen Elizabeth II asked a group of economists, after the financial crash, why they had failed to predict it. There are few moments when the experts who predicted rapid advances in fusion, or genomics, are asked to explain why their estimates were so wrong.[25] Instead, the authority of expertise is protected from scrutiny, and this diminishes our collective ability to learn.

13.3 Science and judgement: how to map and measure what counts as good science and good technology

A practical challenge of wisdom and metacognitive thinking in relation to science is how to judge a new field of science or technology, not just in terms of whether its methods are rigorous and imaginative but also in terms of its potential impacts?[26]

Most technologies are opposed, at least by someone. As Calestous Juma showed, there was opposition to printing the Koran (on the grounds that it lost the authentic aural experience), just as there had been opposition to printing the Bible in vernacular languages (on the grounds that it undermined the vital role of the priests). There was widespread opposition to refrigeration (which threatened an entire ice industry), to railways that despoiled the countryside and to transgenic crops. There will always be someone whose interests or values are threatened. The challenge is to decide how any harms are balanced by any advantages.[27]

Many have tried to find answers, and to suggest more synthetic tools for assessing technologies. There are methods for Constructive Technology Assessment, Anticipatory Assessment, Real-Time Technology Assessment, Value-Sensitive Design and others, and I mentioned earlier some of the methods used in risk assessment, such as those used by the Netherlands Environmental Assessment.

The European Commission promised to do assessments that would back 'responsible innovation', which it defined as an 'approach that anticipates and assesses potential implications and societal expectations . . . with the aim of fostering the design of inclusive and sustainable research and innovation' (though the definition is close to being tautologous, defining responsible innovation as innovation that takes account of its possible effects, without showing how this is to be done). The European Commission also has a large team in its Joint Research Centre who do sophisticated analyses that recognize the complexities of politics and public value.

Each society has to make judgements of the kind that shaped the multiple new rules on artificial intelligence by China in the early 2020s, or whether to accelerate research into autonomous boats or new drugs for particular cancers. But how? And how to avoid being slaves of fashion? David Edgerton, the great historian of technology, comments that 'rockets and atomic power, so beloved in the 1950s and 1960s as world-transforming technologies, are as likely to have made the world poorer rather than richer once all the costs and benefits have been computed' and similar comments may be made a few decades from now about quantum computing or fusion.[28]

The practice of making assessments of this kind has had to find its way without much support from theory. The disciplines that could have provided a more rigorous and useful approach have largely failed to do so. Economics has developed few coherent or comprehensive methods for analysing which kinds of innovation are good and which are bad. It can see when consumers do or don't want to buy something new, and the analysis of externalities can show in retrospect which innovations generate 'bads' (like pollution) as well as benefits (such as cheaper products). But economics offers no ways of doing so ex ante, beyond traditional cost-benefit analysis. Economists from Karl Marx to Erik Brynjolfsson have written about the distributional impact of technology, encouraging a realistic view that technologies tend to hurt some interests and help others. But, on the whole, this has been a minority interest and more concerned with technologies in general, such as automation, than with specific innovations.

Moreover, although many have addressed how individual scientists should think about their own responsibility, and how to integrate ethics and technology,[29] most public funders and innovation agencies use innovation assessment tools that only capture progress towards markets – and the perceptions of potential investors or consumers – or some rough and ready use of cost-benefit analysis, rather than any more comprehensive Technology Assessment, or measurement of potential public or social value.

A reasonable justification for the relative lack of methods is that they are hard: any systematic attempts to assess emerging technologies are fraught with difficulty. No one can predict how any field of science or technology will evolve. Conference speakers love recounting the many examples of people closely involved in key technology sectors – from computing and transport to energy – who dramatically misread how their field would develop.

Another good reason for caution is that even if you can predict how a technology will evolve, it's very hard to predict exactly who will benefit or suffer. In 1867, Marx wrote that the self-acting mule was threatening cotton spinners (a skilled job) and replacing them with unskilled children. Serious industrial experts agreed with him. In fact, the mule spinners weren't put out of jobs – changes in the way factories were managed meant these skilled cotton workers kept their highly paid jobs well into the twentieth century. Forecasts of the effects of technology on jobs over the last fifty years have been equally inaccurate – indeed futurology has consistently exaggerated and misinterpreted the effects of automation on jobs.

Further complicating the task is the difficulty facing anyone trying to define what the counterfactual to any given innovation is. A coalmine despoils nature and emits lots of CO_2. But, if the alternative is to chop down and burn a large forest, the mine might be better both for nature and the climate. This concern applies to other innovations too. If a nuclear war happens at some point, then we would probably be better off had nuclear weapons not been invented. But if you were considering whether to invent the atom bomb in Los Alamos in 1943, your choice wasn't between inventing nuclear weapons or nuclear weapons never existing, but between you inventing them now or someone else inventing them later and, in all likelihood, using them against you.

There are also complex patterns in the economics. As one recent economic analysis argued, 'we could be living in a unique "time of perils", having developed technologies advanced enough to threaten our permanent destruction, but not having grown wealthy enough yet to be willing to spend sufficiently on safety. Accelerating growth during this "time of perils" initially increases risk, but improves the chances of humanity's survival in the long run' yet there are no easily useable ways to analyse exactly how to manage this balance of threats and potential solutions.[30]

These are all reasons for humility. In any system it's useful to allow a fair degree of freedom either for inventors, researchers or entrepreneurs. Excessive application of precautionary principles can inhibit very desirable progress. But it's implausible to conclude that there should be no scrutiny and it is inherently unhealthy for any society to see technologies as things that emerge magically, and over which there is no possibility of control.

A logical alternative is to attempt a staged approach to knowledge, risk and uncertainty. Since it's very hard to know early on what science might become, exploratory research can be relatively untrammelled. Over time, however, the likely impacts become clearer. But this poses a big challenge. When a technology is young, it's hard to assess; when it's mature, it may be too late to reshape it. So, a rough compromise aims early on to explore possibilities and potential threats, and identify potential triggers, or irreversible steps, which could warrant more intensive scrutiny. For almost any technology, the moment of coming near to market, or being purchased for use by a government, should be a trigger. But for others – where the potential risks are particularly large – the critical moment may be an important technical breakthrough, for example, a significant new capability in AI. The key is to keep choices open, not to close them off, or to make revisable decisions which are explicit about what new facts might trigger a rethink.

This becomes even more important as the old divide between basic and applied research breaks down. In many fields the frontiers of science are also the frontiers of engineering: novel ideas in advanced AI, for example, are quickly put into the world. In these cases we need stronger public institutions to investigate, interrogate and clarify what triggers, data or evidence might prompt tighter controls or more investment.

Such analysis can also be helped by recognition that different types of technology have different structures of value. At least five very different kinds of good are sometimes confused or conflated in economics and good policy should be designed to distinguish them.

A first category includes goods with network effects or positive externalities that become more valuable if others are also consuming them – such as telephones and other network technologies. Public health would also fall into this category. Because of their positive externalities, there is a case for judging growth in consumption of these as more valuable to an economy than growth of other kinds of consumption. For similar reasons, innovation that contributes to goods of this kind should be a higher priority for any society.

A second category encompasses more normal commodities like a new material for clothing. Whether or not I consume these doesn't have much impact for better or worse on other people. These are the types of good around which most economics is shaped. Their profitability can be improved by reducing inputs or increasing the extent to which they are reused or recycled. But their external effects are modest.

The third category contains goods that destroy value for some while creating it for others. These include cars (which create pollution, noise and dislocation for those who don't own them), airlines (which disproportionately worsen climate change) and many other industries. Economics recognizes that they produce 'negative externalities'. It measures these when doing exercises in cost-benefit analysis, and policymakers try to internalize them through taxes or regulations. But only the most obvious and material externalities are recognized in economics; and even the ones that are recognized aren't measured in GDP or company accounts.

Fourth, there are what the economist Fred Hirsch called 'positional goods', whose value comes from their exclusivity; these include stately homes and tropical islands developed for luxury tourism, but not many technologies fall into this category; though Concorde, the flagship of UK and French technology policy in the 1960s, was arguably as much about positional value as absolute value.

Finally, there are goods whose very value comes from the harm they cause and the negative externalities created for others. At the extreme are weapons: teenagers buy knives and nations build nuclear missiles or killer robots to frighten others. Their negative impact on lived value is not an unfortunate by-product but rather integral.

It should be obvious that policy and funding should treat these goods very differently. Yet most funding rules make no distinction between them, and this is particularly a flaw of tools such as R&D tax credits. States will have good reasons for wanting some investment in the last category for defence; but they have little reason to want anyone else to benefit from more advanced means of killing people.

So, a first step of assessment involves understanding what kind of good is being created. A next step involves looking in a rounded way at distributional effects of new technologies, even if this is very difficult in advance. This might include potential gains for citizens and consumers (who presumably won't buy the new thing unless they see gains, though in the longer term they may suffer losses, for example, if a new kind of food harms their health); gains (and sometimes losses) to investors; gains to some workers who get better jobs; losses to other workers who lose their jobs or see a cut in pay; and a mix of gains and losses to natural capital.

Such assessments can be comparatively static or dynamic: e.g. will this cluster of technologies create new jobs and business clusters or have knock-on effects on many other sectors in the manner of some general purpose technologies? They're rarely easy because of the sheer number of possible factors involved. But they can at least in principle be reasonably objective and avoid cultural judgements. They can also be done in real time, tracking the actual effects of technologies (e.g. what effects are different national choices over nuclear power and solar having distributionally?)

Next comes ethical assessment, though this too is complex, particularly when new fields of science are emerging. A useful starting point is to look for compatibility with the golden rule: do unto others what you would have them do unto you. Good innovations are ones that we would want for ourselves and those we love. Bad ones clearly breach the golden rule in that the providers would not want themselves to be consumers. Some technologies are clearly compatible with the golden rule while others support predation, making it easier to control, exploit, or conquer. The distinction between different kinds of good suggested earlier makes this obvious. Technologies for war, or for surveillance, are by their very nature contrary to the spirit of the golden rule. There is no missile system or directed energy weapon for which it makes sense for others to do unto you as you would do unto them, and the same is true of computer viruses.

Other technologies are more obviously compatible with the golden rule – like mobile phones that only become valuable if others have them, oral rehydration therapy, yellow fever vaccines, or new crops enriched with vitamins. Others sit in between, like cars that simultaneously provide value to their owners but also take away clean air, space and peace from people who don't have them. The Internet of things with its arrays of sensors is a good example, combining great potential advantages in relation to the efficiency of transport, energy and security, but also risks in relation to misuse of data. An interesting case study was the attempt by Facebook to introduce free Internet access in India,[31] which showed just how different perceptions of value could be.

Then, too, there are technologies of predation that benefit people but leave nature worse off. How you view these depends on just how human-centric your world view is. To some eyes, large-scale mining, whether of the land or the oceans, is by its nature predatory (even when it doesn't come with the messy combination of displacement, abuse, and occasional windfall pay-offs to indigenous communities). To others, it's just the good fortune that humans enjoy thanks to their evolutionary superiority.

Making these kinds of judgement requires wisdom and synthetic thinking and is an aspect of governance in its widest sense, relevant to executive government as well as to parliaments, but also relevant to business, academia, media and civil society. These are the types of assessment that should involve the kinds of science and technology assemblies and technology shaping agencies described in Chapter 10, helping to guide decisions on when to encourage, when to block and when to redirect.

Such assessments do not deliver a single answer, or a single number that can tell us if a field of technology or science is good or bad. Instead, they provide maps, compasses, descriptions of landscapes that make it easier to think, and act, in multiple dimensions.

14

The dialectics of what is and what matters

In the survey for the 2022 global trustworthiness index (based on interviews with tens of thousands of people in 28 countries, run by IPSOS) respondents were asked whether they trusted various roles. Scientists scored a positive of 57 per cent, just below doctors who topped the table, and just above teachers. Others like armed forces, police, judges, priests had lower, but respectable scores. Civil servants were lower again, at 25 per cent, and, right at the bottom, politicians only received a 12 per cent rating. In such a context it's hard for politicians to claim authority to govern or steer science. They are seen as unreliable, dishonest and sometimes corrupt.

Unfortunately, we have no other means of governing. We can surround politics with powerful institutions committed to truth or competence. We can grow media, civil society and education systems that are open, intelligent and concerned with facts. We can encourage scientists to think and act more ethically. But, ultimately, many of the decisions and authority to guide science have to come from politics, if we wish science to be not just our biggest threat but also our strongest ally.

This is why the question of how to accentuate the best and constrain the worst is one of the central political questions of the century. This is the 'science–politics paradox': the paradox that only politics can govern science, but that politics has to change, radically, to be able to do this well, including through learning from science.

There are many answers, but few of them are simple. They need to start with combatting regression: taking on the forces that are anti-science and anti-rational, denying the validity of the scientific method. These forces will include assertions of the unique soul of various nations that promise a deeper truth than mere science, or claims that truth is fractured and infinitely flexible, that the scientists are not the neutral bystanders they pretend to be, and that duties to nation or belief outweigh any duties to scientific truth. In the face of these attacks it's essential to be clear-headed, and willing to fight.

But such battles against regression are not enough. They also need to be combined with battles for progress, and an evolution of institutions that better interweave science and knowledge with power. Extraordinary knowledge demands extraordinary wisdom; yet at the moment we value the first far more than the second. This is the central argument I have made – a three-step argument that embraces and endorses science, then makes the case for its partial subordination to politics, and finally shows how politics itself has to change to make that subordination effective. It is an argument for both politicizing science and scientizing politics, drawing on the ancient view of politics as concerned with how to achieve the good life for the community.

This is a response to the situation I described using Hegel's account of the dialectic of the master and the servant. Here the state was the master that supported science and provided it with funds and prestige. But science grew, empowered by its engagement with the world, until it became in many ways superior to its master who in turn became ever more dependent on it. As a result, the formal arrangements of sovereignty, in which politics remained wholly sovereign, with science just an input, ceased to work.

Science has attempted to reach out to politics, acknowledging its responsibilities, and political structures have welcomed scientific advice. Over the last half-century a lattice of connections has built up: institutions and committees; a myriad of consultations, citizens juries, parliamentary advisory committees and government advisers. All are attempts to integrate the one form of collective intelligence, the intelligence of science, with another, the intelligence of politics.

But these do not go far enough and sometimes miss the point. For example, the spate of initiatives around science ethics are well intentioned but also problematic, since they try to recast political choices as ethical ones.

And so we need a deeper synthesis of science and politics that accounts for the new patterns of shared sovereignty and can help us think, wisely, about the multiple dilemmas of science and technology that are with us already and are only set to intensify, from quantum and genomics to pathogens. The aim must be to improve the metacognition of our societies, our collective ability to think and act in complex environments. This requires more people who are adept at synthesis – people with the training and experience to grasp in their own minds both the scientific and the political dimension of issues. It requires synthetic institutions which combine scientific methods of analysis with political reasoning. And it requires synthetic processes that allow the exploration of the many dimensions of questions and opportunities.

Some of these can be brought into government from science: organizing knowledge commons that bring together the best available knowledge; explicit prediction, and then learning in the light of what happens; testing of ideas, interrogation and experiment. Ensuring that such knowledge commons exist, and that the knowledge they curate is used and useable, should be a core responsibility of the public funders of science.

Some of the answers to the science–politics paradox also lie in what I've described as new logics. These are logics that start with outcomes desired and transitions needed, and work backwards to the available sources of knowledge and power. They are logics that combine attention to what is with attention to what matters. They encourage us to zoom out to the bigger picture of how whole systems work, before zooming back into the practical challenges of change. Such hybrid logics can find homes in new institutions, both within nations and globally, and can draw in resources and prestige and spread in an evolutionary way, through feedback, just as the current dominant logics of politics, science and bureaucracy have done.

These hybrid logics can guide us in answering some of the big questions facing science worldwide. One is the question of alignment: whether brainpower and resources are directed to the issues and problems that matter most (currently they are not). Another is the question of pathways: whether societies have the means to determine and choose which broad pathways they take, whether in food or energy, artificial intelligence or health (too many assume they cannot). A third is the question of productivity: whether the slowdown in productivity in science is faced head-on, and ultimately resolved. A fourth is the question of how to institutionalize truth. Our common culture and democracy cannot survive without a capacity to discover truth and show up falsehood. We need infrastructures of verification to be organized, financed and authorized to block lies and deceptions, whether in politics or in science itself. These more hybrid logics can, in other words, help science and politics to co-evolve, each learning from the other, and they can help us to get a grip on the many examples of governance deserts, from the BSL4 labs I mentioned at the beginning to the frontiers of computing.

Science is the enemy of equilibrium, since new knowledge destabilizes and unsettles. It can be cold and uncomfortable. In this respect politics is closer to everyday life, the lives of humans that are uncertain, unpredictable, except in their finality, which means we live our lives through improvisation and ambiguity and never find final truths, only tools that are useful at some points in time. Science takes the view of the universe. Politics takes the view of people, limited in space and time.

And so, perhaps to restore our own equilibriums, we revert to myth and narrative. Some of those myths simply ignore science, including myths of nationhood. But others tie science into powerful myths. Nations do this through stories of national prowess – conquest, glory and pride. The most common stories of science are still expansive and ambitious, almost like fairy tales: about finding new galaxies, new particles; curing disease; superseding human intelligence. These are still the standard myths, widely spread in many media and shared with schoolchildren. They promise no limits and essentially tap into ideas of discovery that must reach far back in prehistory, to the human urge to wander and find new places to live.

But now we also need other stories as well. We need stories of a science that is responsive, locked in a more intimate relationship with its potential beneficiaries, particularly in fields such as healthcare. We need stories of careful guardianship, particularly of extremely powerful technologies, like nuclear weapons or advanced AI. And we need stories of 'restorative science', which address the very issues science has helped create: climate and ecology; mental health and anxiety; the diseases fuelled by big cities and propelled by travel. These call forth a science that is caring, compassionate, curing and healing, a science that restores equilibrium rather than only breaking it.

Perhaps this is the moral mission for our times, and the moral underpinning of new hybrid logics, with a mission to mobilize our collective intelligence in all its forms for survival and thriving.

Notes

Introduction

1 Activity in these labs is regulated by quite detailed biosafety requirements for handling different organisms. However, the nature and implementation of these rules is very uneven.

2 See for example: this parliamentary debate on the location of such facilities, which includes mention that the 'USGAO has concluded that evidence is lacking that such research can be done safely on the mainland [of the USA]. It cites the outbreak at Pirbright as the best evidence that an island location is preferable . . .': https://publications.parliament.uk/pa/cm200708/cmselect/cmdius/360/360i.pdf

3 For a comprehensive survey of exponential technologies see Azeem Azhar, *Exponential: How Accelerating Technology is Leaving Us Behind and What to Do About It* (Random House Business, 2021).

4 The OECD publishes extensive data on R&D: this is their table for government spending: https://stats.oecd.org/Index.aspx?DataSetCode=GBARD_NABS2007

5 See: the STRINGs report, 2022, which is discussed in more detail later on. Tommaso Ciarli et al., 'Changing Directions: Steering Science, Technology and Innovation towards the Sustainable Development Goals': https://doi.org/10.20919/FSOF1258

6 I've done the same exercise with business leaders and students too: surprisingly few do any better than the politicians.

7 I would sometimes then recommend they read books like Andrew Bloom's *The Tubes: Behind the Scenes at the Internet* (Penguin, 2013).

8 It is perhaps a symptom of the hollowing out of politics that ethics is often used today in a much wider sense to encompass much of what is political.

9 *Nicomachean Ethics*, I.2 (1094 b 4–7).

10 Though, as I show later, politics needs external pressures and restraints as much as any other field: the principle of non-self-sufficiency applies to politics as much as science.

11 Alongside Aristotle's distinction between the good life of the individual and the community, others argue that ethics concerns universal principles, such as the golden rule, whereas morality is embedded in particular communities.

12 An interesting recent article showed this clearly, using the example of the journal *Nature*'s endorsement of Joe Biden: https://www.nature.com/articles/s41562-023-01537-5. It showed that the endorsement did little to increase Biden's support but 'reduced Trump supporters' trust in scientists in general'.

13 'Metacognition is cognition over cognition: the set of higher order cognitive systems that monitor our mental processes . . . [supervising] our learning, evaluating what we know and don't know, whether we are wrong or not . . .' Stanislas Dehaene, *How We Learn: The New Science of Education and the Brain* (Penguin, 2020), p. 193.

14 https://www.sciencecampaign.org.uk/app/uploads/2023/02/CaSE-Public-Opinion-February-2023-Trends-report.pdf

15 I set out a framework for 'anticipatory regulation' in a piece in 2017: https://www.nesta.org.uk/blog/anticipatory-regulation-10-ways-governments-can-better-keep-up-with-fast-changing-industries/. This approach was subsequently turned into new programmes in several countries, such as the 'Regulatory Pioneers Fund' run by the UK government.

1 Uneasy interdependence

1 Very roughly, Plato argued the first, Aristotle the second, though both believed that the best people should rule, with skills most suited to the tasks of governing, and the greatest virtue, and both believed that while a law-based polity was desirable, many tasks could not be predicted by general laws.

2 'In Major Shift, NIH Admits Funding Risky Virus Research in Wuhan | Vanity Fair': https://www.vanityfair.com/news/2021/10/nih-admits-funding-risky-virus-research-in-wuhan

3 A good overview is S. Lewandowsky, P. Jacobs and S. Neil (2022), 'The Lab-Leak Hypothesis Made it Harder for Scientists to Seek the Truth', *Scientific American.*

4 'NIH Says Grantee Failed to Report Experiment in Wuhan That Created a Bat Virus That Made Mice Sicker': https://www.science.org/content/article/nih-says-grantee-failed-report-experiment-wuhan-created-bat-virus-made-mice-sicker

5 https://www.pewresearch.org/science/2020/09/29/science-and-scientists-held-in-high-esteem-across-global-publics/

6 'The Yale Program on Climate Change Communication': https://climatecommunication.yale.edu/
7 Sebastian Levi, 'Publisher Correction: Country-Level Conditions like Prosperity, Democracy, and Regulatory Culture Predict Individual Climate Change Belief', *Communications Earth and Environment* 2, no. 1 (11 March 2021): 1–1: https://doi.org/10.1038/s43247-021-00134-6
8 Bobby Allyn, 'Researchers: Nearly Half of Accounts Tweeting About Coronavirus Are Likely Bots', *NPR*, 20 May 2020, sec. The Coronavirus Crisis: https://www.npr.org/sections/coronavirus-live-updates/2020/05/20/859814085/researchers-nearly-half-of-accounts-tweeting-about-coronavirus-are-likely-bots
9 If one nation, such as China or the US, achieves a significant lead over others.
10 P. Gluckman, 'Policy: The Art of Science Advice to Government', *Nature* 507 (2014): 163–5: https://doi.org/10.1038/507163a
11 For an excellent overview see: D. Mair, L. Smillie, G. La Placa, F. Schwendinger, M. Raykovska, Z. Pasztor and R. van Bavel, 'Understanding our Political Nature: How to Put Knowledge and Reason at the Heart of Political Decision-Making', EUR 29783 EN, Publications Office of the European Union, Luxembourg, 2019, ISBN 978-92-76-08621-5, doi:10.2760/374191, JRC117161
12 For an interesting overview see: Margaret Foster Riley and Richard A. Merrill, 'Regulating Reproductive Genetics: A Review of American Bioethics Commissions and Comparison to the British Human Fertilisation and Embryology Authority', which also concludes that the US could not import the UK model.
13 A recent example of good diagnosis and weak prescription is: 'Presidential Commission for the Study of Bioethical Issues, New Directions: The Ethics of Synthetic Biology and Emerging Technologies'.
14 Tommaso Ciarli et al., 'Changing Directions: Steering Science, Technology and Innovation towards the Sustainable Development Goals' (Brighton, UK: University of Sussex, 20 October 2022): https://doi.org/10.20919/FSOF1258
15 The Sara Cultural Centre and Wood Hotel in Sweden is an example, 20 storeys high.
16 A shift pioneered by UK peer Biban Kidron, who mobilized UN institutions, and the UK parliament, to push the big platforms into taking childhood seriously.
17 L. Lessig, Code: Version 2.0, 4, 2006 and https://cyber.harvard.edu/works/lessig/laws_cyberspace.pdf

18 Ethan Pollock, 'Review of *Science under Socialism in the USSR and Beyond*, by Vera Tolz, Nikolai Krementsov, Paul R. Josephson, Kristie Macrakis, Dieter Hoffmann and Loren Graham', *Contemporary European History* 10, no. 3 (2001): 523–35. http://www.jstor.org/stable/20081809

19 A few decades later it remains unclear whether the claim is accurate.

20 A writer in the Indian paper the Hindu, commented that 'it rankles with us that these impure, beef-eating "materialists", a people lacking in our spiritual refinements, a people whose very claim to civilization we delight in mocking, managed to beat the best of us when it came to nature-knowledge. So, while we hanker after science and pour enormous resources into becoming a "science superpower", we simultaneously devalue its historical and cultural significance and decry its "materialism", its "reductionism" and its "Eurocentrism". We *want* the science of the materialist upstarts from the West but cannot let go of our sense of spiritual superiority.' 'Hindutva's Science Envy – Frontline': https://frontline.thehindu.com/science-and-technology/hindutvas-science-envy/article9049883.ece

21 A popular strain of recent writing – associated with figures such as Stephen Pinker and Sam Harris – argues that Western values are indeed supported by science. I am less convinced and see this as a sophisticated example of confirmation bias. See, for example, James Davison Hunter and Paul Nedelisky, *Science and the Good* (New Haven: Yale University Press, 2018).

22 Naomi Oreskes and Erik M. Conway, *Merchants of Doubt: How a Handful of Scientists Obscured the Truth on Issues from Tobacco Smoke to Global Warming* (London, UK: Bloomsbury Press, 2010).

23 For a good overview of data on food choices and emissions see: https://ourworldindata.org/food-choice-vs-eating-local

24 Adrian Wooldridge, *Measuring the Mind: Education and Psychology in England 1860–1990* is a fascinating account of how liberals and socialists tried to use science to shape the population, often in ways that horrify us now; it argues that the rise of psychology in the late nineteenth century was driven by a desire to control and regulate human behaviour.

25 Christopher Achen and Larry Bartels, *Democracy for Realists* (Princeton, NJ: Princeton University Press, 2016).

26 Martin Gilens and Benjamin I. Page, 'Testing Theories of American Politics: Elites, Interest Groups, and Average Citizens', *Perspectives on Politics* 12, no. 3 (2014): 564.

27 'Genetically Modified Crops: Safety, Benefits, Risks and Global Status |Policy Support and Governance| Food and Agriculture Organization of the United Nations': https://www.fao.org/policy-support/tools-and-publications/resources-details/en/c/1477336/

28 P.J. Crutzen and C. Schwägerl, 2011. 'Living in the Anthropocene: Toward a New Global Ethos' (Yale Environment, 2011), p. 360: http://e360.yale.edu/feature/living_in_the_anthropocene_toward_a_new_glo bal_ethos/2363/

2 What is science and how does it connect to power?

1 In A. Zellner, H. Keuzenkamp and M. McAleer, eds, *Simplicity, Inference and Modelling: Keeping it Sophisticatedly Simple* (Cambridge, UK: Cambridge University Press, 2002).

2 'Sources of Knowledge on the Concept of a Rational Capacity for Knowledge | PDF | Idea | Epistemology', Scribd: https://www.scribd.com/document/482869877/Sources-of-Knowledge-on-the-Concept-of-a-Rational-Capacity-for-Knowledge

3 This distinction remains popular, even though generations of philosophers have shown that the reality of science is much fuzzier.

4 Ray Eames' classic film *The Power of Ten* remains a brilliant introduction to the scales of the universe: https://www.youtube.com/watch?v=0fKBhvDjuy0

5 An interesting strand of the study of science argues against the jump from observation to interpretation. Our concepts – such as the idea of 'the social' are simply too powerful, and too distorting in their effects: better just to describe.

6 The word technology has been quite fluid in its meanings. It used to refer to the academic study of making things (hence institutions like MIT); its more modern meaning of describing the stuff itself dates to the middle of the twentieth century, but now almost anything can be a technology.

7 Helen E. Longino, *Science as Social Knowledge: Values and Objectivity in Scientific Inquiry* (Princeton, NJ: Princeton University Press, 1990): https://doi.org/10.2307/j.ctvx5wbfz

8 See for example: V. Narayanamurti and J. Tsao, *The Genesis of Technoscientific Revolutions* (Cambridge, MA: Harvard University Press, 2021).

9 See for example: David Edgerton, 'From Innovation to Use: Ten Eclectic Theses on the Historiography of Technology', *History and Technology* 16 (1999): 111–36.

10 *The Nature of Technology*, 2009: https://www.simonandschuster.com/books/The-Nature-of-Technology/W-Brian-Arthur/9781416544067

11 Harry Collins, *Gravity's Shadow: The Search for Gravitational Waves* (Chicago, IL: University of Chicago Press, 2004): https://press.uchicago.edu/ucp/books/book/chicago/G/bo3615501.html; James A. Secord, 'Knowledge in Transit', *Isis* 95, no. 4 (2004): 654–72: https://doi.org/10.1086/430657
12 'Kranzberg's Six Laws of Technology – Education, ICT and Philosophy – A Selection of Quotations', 28 March 2019: https://jesperbalslev.dk/kranzbergs-six-laws-of-technology/
13 Michael Mann's *Sources of Social Power* (Cambridge, UK: Cambridge University Press, 1986) is a good summary of the ways in which power is exercised to change other's behaviour. It can be exercised through coercion and force; through economics (paying someone to do something); through political or administrative power of the kind found in hierarchies; and through persuasion or ideology.
14 This is the argument well-made by Jon Agar in his book *Science and Technology in the 20th Century* (Cambridge, UK: Polity, 2010).
15 For one perspective on this see: Bernard Stiegler's 4-volume *Technics and Time*, which argues that this dependence on archival technologies defines much of our civilization, and the political confrontations, for example around platforms.
16 Joseph Glanvill, *The Vanity of Dogmatizing* (1661).
17 Matt Clancy, 'How Common Is Independent Discovery?', Substack newsletter, *What's New Under the Sun* (blog), 22 June 2022: https://mattsclancy.substack.com/p/how-common-is-independent-discovery
18 Dalmeet Singh Chawla, 'Hyperauthorship: Global Projects Spark Surge in Thousand-Author Papers', *Nature*, 13 December 2019: https://doi.org/10.1038/d41586-019-03862-0
19 Aristotle thought that people are by their nature keen to know, and that the most important way of knowing is science, though his conception of science (ἐπιστήμη) is much broader than ours. 'Aristotle, *Metaphysics*, Book 1, Section 980a': http://www.perseus.tufts.edu/hopper/text?doc=Perseus%3Atext%3A1999.01.0052
20 Matt Clancy, 'Innovators Who Immigrate': https://mattsclancy.substack.com/p/innovators-who-immigrate
21 Or, to be more precise, put at the end of a book by his secretary.
22 See: J. Hughes, *The Manhattan Project: Big Science and the Atom Bomb* (London, UK: Icon Books, 2002).
23 Sabine Hossenfelder, 'The World Doesn't Need a New Gigantic Particle

Collider', *Scientific American*: https://www.scientificamerican.com/article/the-world-doesnt-need-a-new-gigantic-particle-collider/

24 Helga Nowotny and Mario Albornoz, *Re-Thinking Science. Knowledge and the Public in an Age of Uncertainty* (Cambridge, UK: Polity, 2003).

25 The argument of Lewis Mumford in multiple books.

26 Paul Josephson 'Science, Ideology and the State' in Mary Jo Nye, ed., *The Cambridge History of Science*, vol. 5, p. 587. See also: Philip Ball, *Serving the Reich: The Struggle for the Soul of Physics under Hitler* (Chicago, IL: University of Chicago Press, 2014).

27 Havi Carel and Ian James Kidd, 'Epistemic Injustice in Healthcare: A Philosophical Analysis', *Medicine, Health Care and Philosophy* 17, no. 4 (November 2014): 529–40: https://doi.org/10.1007/s11019-014-9560-2

28 Peter Burke, *Ignorance: A Global History* (Yale University Press, 2023).

29 Ulrich Binder and Jürgen Oelkers, 'Bildung, Öffentlichkeit und Demokratie im Wandel', in S. Dies, ed., *Der neue Strukturwandel von Öffentlichkeit. Weinheim: Beltz Juventa* (2017), pp. 7–15. Jürgen Habermas, *The Structural Transformation of the Public Sphere: An Inquiry into a Category of Bourgeois Society* (Cambridge, MA: MIT Press, 1989).

30 For an entertaining account of these schemes see Vince Houghton, *Nuking the Moon: And Other Intelligence Schemes and Military Plots Left on the Drawing Board* (Penguin, 2019).

3 The ages of techne and episteme

1 'Innovation is "sudden" and "violent" and, particularly after the French Revolution, is often discussed in terms of "revolt" and what we call revolution today (wars, disorders, schisms), and contrasted with reformation, which is gradual. Innovation is destructive of the social order. This is why innovation is to be feared. The innovator foments a plan to "subvert" things for his own purposes. Innovation may be private as to origin, but it is public with regard to its consequences. Innovation may begin as a small or indifferent thing (adiaphora) but over time it leads to a chain reaction. It creeps imperceptibly, "little by little", into the whole world.' Benoît Godin, 'The English Reformation and the Invention of Innovation, 1548–1649', *Contributions to the History of Concepts* 17, no. 1 (1 June 2022): 1–22: https://doi.org/10.3167/choc.2022.170101

2 Joseph Ben-David and Gad Freudenthal, eds, *Scientific Growth: Essays on the Social Organization and Ethos of Science* (University of California Press, 1991), p. 339.

3 Jurgen Renn, *The Evolution of Knowledge* (Princeton, NJ: Princeton University Press, 2020), p. 218.

4 M. Polanyi, *Personal Knowledge: Towards a Post-Critical Philosophy* (Chicago, IL: University of Chicago Press, 1962), p. 182, referring to the formal propositional knowledge of science.

5 P. Bowler and I. Morus, 'The Organisation of Science' in *Making Modern Science: A Historical Survey* (Chicago: University of Chicago Press, 2005).

6 As Richard Nelson pointed out in his book *The Moon and the Ghetto* 'it may be enormously more difficult to design policies to equalize educational achievement or to eliminate prejudices, than to design spacecraft to go to the moon'. He then commented 'for truly intractable problems the most we can expect from rational analysis is understanding which deters us from trying costly remedies that cannot work'. But, as I've pointed out elsewhere, this turned out to be wrong. Many of the problems thought to be intractable at that time, such as overpopulation, went on to be substantially solved. Richard Nelson, *The Moon and the Ghetto*, 1977.

7 'Progress against Cancer? | NEJM': https://www.nejm.org/doi/full/10.1056/NEJM198605083141905

8 Surh Young-Joon, 'The 50-Year War on Cancer Revisited: Should We Continue to Fight the Enemy Within?', *Journal of Cancer Prevention* 26, no. 4 (30 December 2021): 219–23: https://doi.org/10.15430/JCP.2021.26.4.219

9 Loren Graham, 'Science in the New Russia', *Issues in Science and Technology* 19, no. 4 (2003).

10 Q. Schiermeier, 'Russia Aims to Revive Science after Era of Stagnation', *Nature* 579, 332–6 (2020): doi: https://doi.org/10.1038/d41586-020-00753-7

11 The new funding model remains very small compared to the conventional model. Foreign funding of science in Russia accounts for less than three per cent of the country's total science spending. Russia's first private science foundation was shut down in 2015 after being designated as a 'foreign agent' and scientists are increasingly asked to have their papers checked by the Federal Security Service before publication. Despite Russia having the fourth largest scientific workforce in the world, the number of scientists has fallen nearly threefold since 1990. A third of researchers are above the retirement age and the average is over fifty (in the UK 76% of scientists are under thirty).

12 Sanei Mansooreh, 'Human Embryonic Stem Cell Science and Policy: The Case of Iran', *Social Science and Medicine* 98 (December 2013): 345–50.

13 Michael Neufeld, 'The Guided Missile and the Third Reich: Peenemünde and the Coming of the Ballistic Missile Era', in Monika Renneberg and Mark Walker, eds, *Science, Technology and National Socialism* (Cambridge, UK: Cambridge University Press, 1994), pp. 51–71.

14 E. Geissler. 'Biological Warfare Activities in Germany, 1923–45', in E. Geissler and J.E. van Courtland Moon, eds, *Biological and Toxin Weapons: Research, Development, and Use from the Middle Ages to 1945* (Oxford, UK: Oxford University Press, 1999), pp. 91–126.

15 See for example: Mu-ming Poo, 'Towards Brain-Inspired Artificial Intelligence', *National Science Review* 5, issue 6 (November 2018): 785: https://doi.org/10.1093/nsr/nwy120

16 https://www.nasa.gov/commercial-orbital-transportation-services-cots

17 Nur Ahmed, 'The Growing Influence of Industry In AI Research' (with N. Thompson and M. Wahed), *Science* (2023).

18 Shawn Kantor and Alexander Whalley, 'Moonshot: Public R&D and Growth', *NBER* (April 2022).

19 https://www.aps.org/publications/apsnews/201708/backpage.cfm

20 G. Agamben, *The Kingdom and the Glory* (Stanford, CA: Stanford University Press, 2011).

21 Quoted in *A History of the Modern Fact*, p. 132.

22 Michel Foucault, *On the Government of the Living: Lectures at the Collège de France, 1979–1980*, ed. Michel Senellart, trans. Graham Burchell (Basingstoke, New York: Palgrave Macmillan, 2014).

23 Ian Hacking, 'Trial by number', *Science*, 84 (1984): 67–70.

24 James Scott, *Seeing Like a State* (Newhaven, CT: Yale University Press, 1998).

25 Laurence E. Lynn, Jr, ' Public Management: A Brief History of the Field' in E. Ferlie, L. Lynn and C. Pollitt, eds, *Handbook of Public Management* (Oxford, UK: Oxford University Press, forthcoming). Frederick William I had started training officials and established two university chairs in administrative subjects in the 1720s (by the end of the eighteenth century there were 23 such chairs), while Frederick the Great instituted examinations and a civil service commission.

26 *A History of the Modern Fact*, p. 308.

4 Science bites back

1 S. Jasanoff, 'Technologies of Humility', *Nature*, 450 (2007): 7166; O. Renn, 'Three Decades of Risk Research', *Journal of Risk Research* 1, no. 1 (1998): 49–71.

2 See Charles Jones, 'Life and Growth', *Journal of Political Economy*, 124 (2) (2016): 539–78: http://dx.doi.org/10.1086/684750

3 Ortwin Renn, *Risk Governance: Coping with Uncertainty in a Complex World, Earthscan* (London: Routledge, 2008).

4 Jeanne Guillemin, 'Scientists and the History of Biological Weapons: A Brief Historical Overview of the Development of Biological Weapons in the Twentieth Century', *EMBO Reports* 7, no. Spec. No. (July 2006): S45–49: https://doi.org/10.1038/sj.embor.7400689

5 Linsley and Shrives: https://pure.york.ac.uk/portal/en/publications/risk-reporting-a-study-of-risk-disclosures-in-the-annual-reports-

6 'Odds of Dying', *Injury Facts* (blog): https://injuryfacts.nsc.org/all-injuries/preventable-death-overview/odds-of-dying/

7 Mary Douglas and Aaron Wildavsky, *Risk and Culture: An Essay on the Selection of Technological and Environmental Dangers*, 1st edn (University of California Press, 1982): https://www.jstor.org/stable/10.1525/j.ctt7zw3mr

8 Tom Stafford, 'Throwing Science at Anti-Vaxxers Just Makes Them More Hardline', *The Conversation* (19 February 2015): http://theconversation.com/throwing-science-at-anti-vaxxers-just-makes-them-more-hardline-37721

9 Karen Stenner, 'The Authoritarian Dynamic', *Cambridge Studies in Public Opinion and Political Psychology* (Cambridge, UK: Cambridge University Press, 2005): https://doi.org/10.1017/CBO9780511614712

10 This expectation may vary by culture. It seems to be less of an expectation in the USA, for example.

11 The decisive book on the prevalence of risk is U. Beck, *Risk Society: Towards a New Modernity* (London, UK: Sage, 1992).

12 *The Politics of Uncertainty: Challenges of Transformation* (Routledge and CRC Press): https://www.routledge.com/The-Politics-of-Uncertainty-Challenges-of-Transformation/Scoones-Stirling/p/book/9780367903350

13 https://www.pbl.nl/en/publications/guidance-for-uncertainty-assessment-and-communication

14 Tobias Olofsson et al., 'The Making of a Swedish Strategy: How Organizational Culture Shaped the Public Health Agency's Pandemic Response', *SSM – Qualitative Research in Health* 2 (1 December 2022): 100082: https://doi.org/10.1016/j.ssmqr.2022.100082

15 'Risk Savvy, by Gerd Gigerenzer': https://www.mindtherisk.com/literature/67-risk-savvy-by-gerd-gigerenzer

16 The original Cassandra, daughter of Priam, had the gift of accurate prophecy but was fated never to be believed.

17 The UK now has an extensive ecosystem for risk assessments. These include a National Risk Register and National Security Risk Assessment; insurance risk assessments (e.g. for university labs and industry); global measures such as GHSI: https://www.ghsindex.org/, which measures biosafety and also dual-use research and culture of responsible science: https://www.ghsindex.org/wp-content/uploads/2021/12/United-Kingdom.pdf

18 See: Ulrich Beck, *Risk Society: Towards a New Modernity*. There is now a large literature on the anthropology and cultures of risk. A classic study on the Space Shuttle Challenger disaster, for example, showed that the bureaucratic hope that precedents, guidelines and rules could make risk manageable, in this case with the 'nested cultural scripts' of the engineering profession, NASA, and the Marshall Space Flight Center, and encouraged decision-makers who were working with uncertain and dangerous systems to act as if they were not.

19 https://www.weforum.org/agenda/2019/11/countries-preparedness-pandemics/

20 'The Limits to Growth+50', Club of Rome: https://www.clubofrome.org/ltg50/

21 Filippa Lentzos, Michael S. Goodman and James M. Wilson, 'Health Security Intelligence: Engaging across Disciplines and Sectors', *Intelligence and National Security* 35, no. 4 (6 June 2020): 465–76: https://doi.org/10.1080/02684527.2020.1750166

22 For example, the availability of Kilobaser DNA & RNA Synthesizers on sale for a few tens of thousands of dollars seemed to suggest mass access to tools that could cause huge numbers of deaths: https://kilobaser.com/dna-and-rna-synthesizer/

23 Lentzos, Goodman and Wilson, op. cit.

24 John M. Barry, 'The Next Pandemic', *World Policy Journal* 27, no. 2, MIT Press (Summer 2010).

25 Jessica Weinkle and Roger Pielke, 'The Truthiness about Hurricane Catastrophe Models' (2017): https://journals.sagepub.com/doi/abs/10.1177/0162243916671201?journalCode=sthd

26 'Cell Phones and Cancer Risk Fact Sheet – NCI', cgvArticle, 3 October 2022, nciglobal,ncienterprise: https://www.cancer.gov/about-cancer/causes-prevention/risk/radiation/cell-phones-fact-sheet

27 Alan Irwin and Brian Wynne, eds, *Misunderstanding Science? The Public Reconstruction of Science and Technology* (Cambridge, UK: Cambridge University Press, 1996), p. 75.

28 Philip M. Linsley and Philip J. Shrives, 'Mary Douglas, Risk and Accounting Failures', *Critical Perspectives on Accounting* 20, no. 4 (1 May 2009): 492–508: https://doi.org/10.1016/j.cpa.2008.05.004

29 Linsley and Shrives, ibid.

30 Niheer Dasandi et al., 'Positive, Global, and Health or Environment Framing Bolsters Public Support for Climate Policies', *Communications Earth and Environment* 3, no. 1 (20 October 2022): 1–9: https://doi.org/10.1038/s43247-022-00571-x

31 Jason Wei et al., 'Emergent Abilities of Large Language Models' (arXiv, 26 October 2022): https://doi.org/10.48550/arXiv.2206.07682

32 R.J. Lempert, S.W. Popper and S.C. Bankes, 'Shaping the Next One Hundred Years: New Methods For Quantitative, Long-Term Policy Analysis', RAND (2003), 10.1016/j.techfore.2003.09.006: https://www.sciencedirect.com/science/article/pii/S0040162521003711

5 *The scientific view of politics as corruptor*

1 Joseph Ben-David and Gad Freudenthal, eds, *Scientific Growth: Essays on the Social Organization and Ethos of Science* (University of California Press, 1991), p. 339.

2 https://philarchive.org/archive/BROMCA-5

3 Reiner Grundmann, 'Knowledge Politics', in *The Blackwell Encyclopedia of Sociology* (Wiley, 2018) and *The Problem of Expertise in Knowledge Societies* (Minerva, 2016).

4 Isabelle Stengers, 'Another Science Is Possible', *Perlego*: https://www.perlego.com/book/1536458/another-science-is-possible-a-manifesto-for-slow-science-pdf

5 Parallel arguments can be found in the writings of Peter Godfrey-Smith, *Theory and Reality: An Introduction to the Philosophy of Science* (Chicago, IL: University of Chicago Press, 2003), which is also particularly good on the difficulty of proving many heuristics used in everyday science (of which this is one).

6 Robert K. Merton, 'The Normative Structure of Science', in *The Sociology of Science: Theoretical and Empirical Investigations* (Chicago, IL: University of Chicago Press, 1979 [1942]).

7 See also, Robert K. Merton, *Social Theory and Social Structure* (New York: Free Press, 1957), pp. 574–85. Merton thought that modern science built on theology: 'Faith in the possibility of science' drew on the theological emphasis on discovering God in his creation, and 'an emphasis upon experience and reason as bases for action and belief . . .'

8 Michael Polanyi, *The Scientific Monthly* 60, no. 2 (Feb., 1945): 141–50.

9 Sometimes this required institutional innovation, as when scientists tried to bridge geopolitical conflicts through the International Union of Pure and Applied Chemistry (IUPAC) and the International Union of Pure and Applied Physics (IUPAP), creating a joint working group to end controversy over the discovery of transfermium elements – those elements with an atomic number greater than 100: Audra J. Wolfe, *Freedom's Laboratory*, 2020: https://doi.org/10.1353/book.99572

10 Guillemin, 'Scientists and the History of Biological Weapons'.

11 For a good overview of the field see: Wendell Wallach and Anja Kaspersen, 'Creative Reflections on the History and Role of AI Ethics', Carnegie Council, Artificial Intelligence and Equality Initiative, 26 May 2021: www.carnegiecouncil.org/media/series/aiei/20210426-creative-reflections-history-role-artificial-intelligence-ethics-wendell-wallach

12 69% of US adults would support a six-month ban on some AI development, while 13% oppose it: https://today.yougov.com/topics/technology/survey-results/daily/2023/04/03/ad825/2

13 https://www.trinityhouse.co.uk/

14 https://www.ul.com/about/history. An example was the problem of warehouse spills of cleaning products with proprietary formulas. Warehouse owners couldn't just wash the spills down the drain, since that might violate EPA water regulations. But product providers didn't want to share ingredients (proprietary). UL developed an app that could read the barcode of a product and tell the owner of the warehouse what to do to comply with EPA rules without sharing the underlying formula.

15 Evan S. Michelson, 'Philanthropy and the Future of Science and Technology', *Perlego*: https://www.perlego.com/book/1583980/philanthropy-and-the-future-of-science-and-technology-pdf

16 As, for example, in the writings of the philosopher William MacAskill.

17 Henry Etzkowitz and Loet Leydesdorff, 'A Triple Helix of University – Industry – Government Relations: Introduction', *Industry and Higher*

Education 12, no. 4 (1 August 1998): 197–201: https://doi.org/10.1177/09 50421229801200402

18 Chris Freeman, 'The "National System of Innovation" in Historical Perspective', *Cambridge Journal of Economics* 19, no. 1 (1995): 5–24.

19 Peter Evans, *Embedded Autonomy: States and Industrial Transformation* (Princeton, NJ: Princeton University Press, 1995). Evans suggested that scientists are embedded in social ties that bind the state to society and provide institutionalized channels for the continual negotiation and renegotiation of goals and policies, in part linked to the ways different nations navigated their different niches in the global division of labour. See also: Tim Flink and David Kaldewey, 'The New Production of Legitimacy: STI Policy Discourses beyond the Contract Metaphor', *Research Policy* 47, no. 1 (2018): 14–22.

20 Tim Flink and David Kaldewey, 'The New Production of Legitimacy'.

21 For example, in the work of Jurgen Habermas and Herbert Marcuse – with biases further encoded into technology.

22 Langdon Winner, *The Whale and the Reactor: A Search for Limits in an Age of High Technology* (Chicago, IL: University of Chicago Press); Donald MacKenzie and Judy Wacjman, *The Social Shaping of Technology* (Open University Press, 1985); Eric Schatzberg, *Technology: Critical History of a Concept* (Chicago, IL: University of Chicago Press, 2018).

23 Donna Haraway, 'Situated Knowledge: The Science Question in Feminism as a Site of Discourse on the Privilege of Partial Perspective', *Feminist Studies*, 1988, 14: 575–99.

24 Evelyn Fox Keller, 'Reflections on Gender and Science', in the *Cambridge Companion to Feminism in Philosophy* (Cambridge, UK: Cambridge University Press, 2000).

25 Figures like Trevor Pinch, Thomas Hughes, Steven Shapin and many others.

26 A related theme was interest in what for a time was called post-normal science: Jerome R. Ravetz, *Scientific Knowledge and its Social Problems* (Oxford, UK: Oxford University Press, 1979); Jerome R. Ravetz, 'Usable Knowledge, Usable Ignorance: Incomplete Science with Policy Implications', *Knowledge* 9 (1) (September 1987): 87–116: doi:10.1177/107554708700900104.S2CID 146551904.

27 Joseph Harris, 'Science and Democracy Reconsidered', *Engaging Science, Technology, and Society* 6 (2020): 102–10.

28 *The Song of the Cell*, 2022: https://www.simonandschuster.com/books/The-Song-of-the-Cell/Siddhartha-Mukherjee/9781982117351

29 Alvin M. Weinberg, 'Science and Trans-Science', *Minerva* 10, no. 2 (1972): 209–22.

30 'Bruno Latour, the Post-Truth Philosopher, Mounts a Defense of Science', *New York Times*: https://www.nytimes.com/2018/10/25/magazine/bruno-latour-post-truth-philosopher-science.html

31 Latour also realized that his arguments had become problematic in an era of anti-science sentiments: he argued that scientists needed to 'regain some of the authority of science. That is completely opposite from where we started doing science studies', in Jop de Vrieze, 'Bruno Latour, a Veteran of the "Science Wars" Has a New Mission', *Science* 10 (October 2017).

32 'W.I. Thomas on Social Organization and Social Personality', *Selected Papers by Morris Janowitz* 1966: 301.

33 See: Pierre Bourdieu, *The Science of Science and Reflexivity* (Cambridge, UK: Polity, 2004), p. 9. His work has aged well, though his personal animus against Latour is unfortunate.

6 Master, servant and multiple truths

1 'Faith, Certainty and the Presidency of George W. Bush', *New York Times*: https://www.nytimes.com/2004/10/17/magazine/faith-certainty-and-the-presidency-of-george-w-bush.html

2 One of the odder experiences of my life was spending a week in the Kremlin with Putin's team, just after he came to power.

3 He echoed Plato's advocacy of the noble lie, that 'would have a good effect, making [the people] more inclined to care for the state and one another'.

4 See: Jon Agar, op. cit.

5 *Барсенков А. С., Вдовин А. И.* История России. 1917–2007 – М.: Аспект Пресс, 2008. – С. 418.

6 https://www.rathenau.nl/en/science-figures/policy-and-structure/infrastructure-knowledge/science-policy-and-innovation-policy

7 I teach the science of public administration but have to confess that it is a long way from being a science in the full sense.

8 Charles-Louis de Secondat, 'Montesquieu, the Spirit of Laws (1748)', n.d.

9 The history of large language models began in 2018, when Google showed an impressive prototype called BERT (Bidirectional Encoder Representations from Transformers), that could generate plausible prose in answer to questions. A year later, in 2019, Open AI released GPT-2, a vast model with 1.5bn parameters, succeeded by GPT-3 (with 175bn parameters), then ChatGPT

which, in 2022, became the fastest adopted software in history with 100m users in its first month, and which could write plausible speeches, press releases and summaries for ministers, and in 2023 was followed by GPT4.

7 Clashing logics

1 One recent writer suggested distinguishing types of knowledge: a 'cognitive, mental or "internal" dimension; a material, embodying or "external" dimension; and a social dimension, referring to the societal processes involved in production, sharing, transmitting and appropriating knowledge'. Jurgen Renn in *The Evolution of Knowledge.*

2 https://www.sciencedirect.com/science/articl',abs/pii/S0304422X21000589

3 C.W. Mills, 'Language, Logic, and Culture', *American Sociological Review*, 4(5) (1939): 670–80.

4 Pierre Bourdieu, *Distinction: A Social Critique of the Judgement of Taste*, 11, print (Cambridge, MA: Harvard University Press, 2002).

5 See also: Lauren Valentino, 'Cultural Logics: Toward Theory and Measurement', *Poetics*, Measure Mohr Culture, 88 (1 October 2021): 101574: https://doi.org/10.1016/j.poetic.2021.101574

6 G. Simmel, *The View of Life* (Chicago, IL: University of Chicago Press, 2010), p. 37 (I have slightly paraphrased his words).

7 Simmel, op. cit., p. 164 (again slightly paraphrased).

8 Mukherjee, *Song of the Cell.*

9 V. Narayanamurti and J. Tsao, *The Genesis of Technoscientific Revolutions* (Cambridge, MA: Harvard University Press, 2021), p. 150.

10 In French the word 'politique' applies to both policies – the aims and programmes of groups or states – and the world of politics, debate and competition for power. Here I focus on the second.

11 N. Luhmann, *The Politics of the Welfare State*, p. 111.

12 John E. Elliott, 'Joseph A. Schumpeter and the Theory of Democracy', *Review of Social Economy* 52, no. 4 (1994): 280–300.

13 See Paolo Quattrone's fascinating piece: https://aeon.co/essays/lessons-in-corporate-governance-from-the-jesuits

14 Max Weber, 'Politics as a Vocation', *The Pacific Sociological Review* 24, no. 1 (1981): 17–44.

15 'Carl Schmitt's Concept of the Political': https://www.jstor.org/stable/2127676

16 Indeed, many political systems focus much more on what they can achieve for themselves in the here and now, rather than fixating on enemies, whether real or

imagined. As one more recent theorist put it, an ideal system may be one 'that produces a broader representation of the general interest, while also respecting the importance for each district of being represented by someone who knows its situation and needs'. Jon Elster, *Securities against Misrule: Juries, Assemblies, Elections* (Cambridge; New York: Cambridge University Press, 2013).

17 'Innovation and Its Enemies: Why People Resist New Technologies', Weatherhead Center for International Affairs: https://wcfia.harvard.edu/publications/innovation-and-its-enemies-why-people-resist-new-technologies

18 As one recent work of political science argued 'the idealistic justification of democracy as human rationality in pursuit of the common good serves only too well to provide cover for those who profit from the distortions and biases in the policy-making processes of actual democracies'. Christopher Achen and Larry Bartels, 'Democratic Ideals and Realities', in *Democratic Ideals and Realities* (Princeton, NJ: Princeton University Press, 2017), pp. 1–20: https://doi.org/10.1515/9781400888740-003

19 This is an observation made by Ibn Khaldun some eight hundred years ago. I wrote about his work at length in my book *Good and Bad Power* (Penguin, 2006).

20 Good books on bureaucracy include James Q. Wilson, *Bureaucracy* (Basic Books 1991) and Edward Page, *Policy Without Politicians: Bureaucratic Influence in Comparative Perspective* (Oxford, UK: Oxford University Press, 2012).

21 An excellent recent study of China showed through detailed analysis of thousands of directives how bureaucracies try to direct a society. The directives varied from 'black' (scaling up or accelerating something) through 'red' (stopping something) to 'grey' (deliberately ambiguous but with a direction). Such 'grey' directives give space for exploration and innovation and, according to the analysis, new industries like e-commerce, big data, and artificial intelligence, attract the highest proportion of grey directives, as the government tries to 'cross the river by feeling the stones'. Yuen Yuen Ang, 'How Beijing Commands: Grey, Black, and Red Directives from Deng to Xi', *The China Quarterly* (26 August 2022). SSRN: https://ssrn.com/abstract=4201534

22 H. Arendt, *Origins of Totalitarianism.*

23 W.E. Bijker, T.P. Hughes and T.J. Pinch, eds, *The Social Construction of Technological Systems* (Cambridge, MA: MIT Press, 1989), p. 68.

24 James March and Johan Olsen suggested that two different types of logic can be found among people who work in institutions, which includes the great majority of scientists, politicians and bureaucrats. One is a logic of

appropriateness, based on rules, which are derived from roles and identities. If you follow the rules, you are doing your job well, and if you breach them, you are culpable. The second is a logic of consequentiality, based on one's interests or preferences and derived from a 'calculation of consequences and expected utility'. The authors suggested that while the logic of appropriateness is likely to motivate actors in institutional settings, actors use the logic of consequentiality to explain those actions: 'different logics of action are used for different purposes, such as making policies and justifying policies'. So, the scientific logic may look different when facing inwards and outwards: facing inwards it may be more concerned with appropriate rules of scientific method, while in facing outwards it's necessary to make claims about results and impacts in the world. Similarly, the bureaucrat may discuss choices with other bureaucrats in terms of rules and appropriateness but, when talking to anyone else, has to talk in terms of consequences. James G. March and Johan P. Olsen 'Institutional Perspectives on Political Institutions', *Governance* 9, no. 3 (1996): 247–64.

25 William MacAskill, 'Opinion: The Case for Longtermism', *New York Times*, 5 August 2022, sec. Opinion: https://www.nytimes.com/2022/08/05/opinion/the-case-for-longtermism.html

26 This is part of their appeal to politicians. They can gain the credit for launching missions, safe in the knowledge that they will probably be long gone by the time the mission is quietly declared a failure.

27 W. Sellars, *Science, Perception and Reality* (London: Routledge and Kegan Paul; and New York: The Humanities Press, 1963).

28 'Introduction to the Study of the Law of the Constitution (LF Ed.) | Online Library of Liberty': https://oll.libertyfund.org/title/michener-introduction-to-the-study-of-the-law-of-the-constitution-lf-ed

29 Pierre Bourdieu, echoing Goffmann, in *The Science of Science and Reflexivity*, (Cambridge, UK: Polity, 2004), p. 9.

30 There is a long-standing debate in social science about how much the social world is objectively made up of facts, and how much it is made up of perceptions. Attention to the role of perceptions emanates from Schopenhauer's claim that the social world is just a representation, a reflection of our will.

8 Split sovereignty, or the role of knowledge in corroding the supremacy of politics

1 Pierre Charbonnier, *Affluence and Freedom: An Environmental History of Political Ideas* (Cambridge, UK: Polity, 2021) comments cogently on the interaction of political ideas of freedom and sovereignty and the natural world.

2 In any hierarchy there is an apex that is partly inexplicable – where the describable becomes opaque: that is as true of the board and the CEO as it is of the President, and perhaps something in our evolutionary history and all of our histories as children make us want to believe in the mystery and magic of leaders, to believe that they understand the world we are in.

3 Articles 10 and 71–4.

4 The critics thought it went too far, prescribing content rather than principles, echoing the views of figures like Friedrich Hayek, who argued that it was an 'abuse' of 'democracy' if one tries 'to give it a substantive content prescribing what the aim of those activities ought to be'.

9 Democracy meets science

1 As pulled together by an advisory committee of scientists and fed through the GCSA and CMO to ministers.

2 Richard Horton: 'It's the Biggest Science Policy Failure in a Generation', *Financial Times*: https://www.ft.com/content/8e54c36a-8311-11ea-b872-8db45d5f6714

3 For a very thorough, critical and well-informed assessment see: 'Lessons from the UK's Handling of Covid-19 for the Future of Scientific Advice to Government: A Contribution to the UK Covid-19 Public Inquiry'.

4 Justus Lentsch and Peter Weingart, eds, *The Politics of Scientific Advice: Institutional Design for Quality Assurance* (Cambridge; New York: Cambridge University Press, 2011). See also my essay in James Wilsdon and Robert Doubleday, eds, *Future Directions for Scientific Advice in Whitehall*, Centre for Science and Policy, Cambridge, 2013.

5 https://eptanetwork.org/

6 https://english.wrr.nl/

7 Sheila Jasanoff, 'Serviceable Truths: Science for Action in Law and Policy', *Texas Law Review* 93 (n.d.). The most successful boundary spanners link 'the ordering of nature through science and technology and the ordering of society through power and culture . . . in a process of coproduction that confirms that 'the ways in which we know and represent the world' are 'inseparable from the ways in

which we choose to live in it'. Sheila Jasanoff, *Science at the Bar: Law, Science, and Technology in America* (Cambridge, MA: Harvard University Press, 1995).

8 Ed Yong, 'What Even Counts as Science Writing Anymore?', *The Atlantic* (2 October 2021): https://www.theatlantic.com/science/archive/2021/10/how-pandemic-changed-science-writing/620271/

9 Sheila Jasanoff, *The Fifth Branch* (Cambridge, MA: Harvard University Press): https://www.hup.harvard.edu/catalog.php?isbn=9780674300620

10 Lentsch and Weingart, *The Politics of Scientific Advice.*

11 He was also right to warn against the many dark arts used by lobbies of all kinds to discredit their enemies and was himself on the receiving end of public attacks by the President's science adviser and concerted attempts to cut off his funding.

12 Bruce L.R. Smith and Roger A. Pielke, 'Can Science Policy Advice Be Disinterested?', *Issues in Science and Technology* 24, no. 4 (2008): 92–5.

13 Brian W. Head, 'Evidence-Based Policymaking – Speaking Truth to Power?' (5 November 2013), p. 297: https://doi.org/10.1111/1467-8500.12037

14 E.C. McNie, A. Parris and D. Sarewitz, 'Improving the Public Value of Science: A Typology to Inform Discussion, Design and Implementation of Research, *Research Policy*, 45, no. 4 (2016): 884–95.

15 *Making Sense of Science for Policy under Conditions of Complexity and Uncertainty* (DE: Science Advice for Policy by European Academies, 2019): https://doi.org/10.26356/masos

16 Elias G. Carayannis and David F.J. Campbell, '"Mode 3" and "Quadruple Helix": Toward a 21st Century Fractal Innovation Ecosystem', *International Journal of Technology Management* 46, nos. 3–4 (January 2009): 3: https://doi.org/10.1504/IJTM.2009.023374

17 David L. Sackett et al., 'Evidence Based Medicine: What It Is and What It Isn't', *BMJ* 312, no. 7023 (13 January 1996): 71–2: https://doi.org/10.1136/bmj.312.7023.71

18 D. Collingridge, *The Social Control of Technology* (St Martin's Press, 1982).

19 'Pre-Print: What Works to Promote Research-Policy Engagement?': https://transforming-evidence.org/resources/what-works-to-promote-research-policy-engagement

20 https://journals.sagepub.com/stoken/default+domain/10.1177/1529100618772271-free/full

21 '"Opening Up" and "Closing Down": Power, Participation, and Pluralism in the Social Appraisal of Technology – Andy Stirling, 2008', https://journals.sagepub.com/doi/10.1177/0162243907311265

22 Donald Schön's *The Reflective Practitioner: How Professionals Think in Action* (1983) for example was an important book in the tradition of US pragmatism attending to how professions combined knowledge, experience, reflection, dialogue and improvisation to solve difficult problems.
23 Treasury Board of Canada Secretariat, 'Experimentation Works', 24 June 2019: https://www.canada.ca/en/government/publicservice/modernizing/experimentation-works.html
24 Jasanoff, 'Serviceable Truths: Science for Action in Law and Policy'.
25 Continuous sharing of knowledge is critical to making these work, so that instead of periodic inquiries or White Papers, adaptation can be continuous and, if necessary, very fast. Dani Rodrik and Charles Sabel, 'Building a Good Jobs Economy', in Danielle Allen, Yochai Benkler, Leah Downey, Rebecca Henderson and Josh Simons, eds, *A Political Economy of Justice* (Chicago, IL: University of Chicago Press, 2022), pp. 61–95.
26 These, and the challenges of delegation, are well explored in David Guston, *Between Politics and Science*, 2000.
27 Ben Almassi, 'Relationally Responsive Expert Trustworthiness', *Social Epistemology* 36, no. 5 (3 September 2022): 576–85: https://doi.org/10.1080/02691728.2022.2103475
28 'Inside the Suspicion Engine', investigation by *Lighthouse* and *Wired Magazine*, 6 March 2023.
29 AI is likely to become evermore part of everyday decision-making and it can provide a more neutral balance to often flawed human decision-making. But, as Sir David Spiegelhalter, former president of the Royal Statistical Society, commented: 'there is too much hype and mystery surrounding machine learning and algorithms . . . [municipalities] should demand trustworthy and transparent explanations of how any system works, why it comes to specific conclusions about individuals, whether it is fair, and whether it will actually help in practice'. We cannot help depending on systems and machines that we cannot understand. But we can put in place systems that will help us ask the right questions and hold power to account.
30 Ben Almassi, 'Relationally Responsive Expert Trustworthiness', *Social Epistemology* 36, no. 5 (2022): 576–85: DOI: 10.1080/02691728.2022.2103475
31 Kyle Whyte and Robert Crease 'Trust, Expertise, and the Philosophy of Science', *Synthese* 177, no. 3 (2010): p. 423.

32 See for example L. Schiebinger et al., eds, 'Gendered Innovations in Science, Health and Medicine, Engineering and Environment' (2011–2018): http://gen deredinnovations.stanford.edu/index.html

33 J. Street, K. Duszynski, S. Krawczyk and A. Braunack-Mayer, 'The Use of Citizens' Juries in Health Policy Decision-Making: A Systematic Review', *Soc. Sci. Med.* (May 2014): 109: 1–9. doi: 10.1016/j.socscimed.2014.03.005. Epub 6 March 2014. PMID: 24657639

34 J.S. Dryzek, A. Bachtiger, S. Chambers, J. Cohen, J.N. Druckman, A. Felicetti, J.S. Fishkin, D.M. Farrell, A. Fung, A. Gutmann, H. Landemore, J. Mansbridge, S. Marien, M.A. Neblo, S. Niemeyer, M. Setala, R. Slothuus, J. Suiter, D. Thompson and M.E. Warren, 'The Crisis of Democracy and the Science of Deliberation', *Science* (New York, 2019); 363 (6432): 1144–6.

35 Maximilian Fries, 'Online Conference: "Green Transition of the Chemical Industry" with the EU Environment Commissioner on Tue, 1/9, 10–12:30', Sven Giegold – Mitglied der Grünen Fraktion im Europaparlament, 24 August 2020: https://sven-giegold.de/en/online-conference-green-transition-chemic al-industry/

36 https://sciencewise.org.uk/projects-and-impacts/

37 https://www.science.org/doi/full/10.1126/science.299.5609.977

38 An excellent book on these topics is Mark Brown, *Science and Democracy* (Cambridge, MA: MIT Press 2009).

39 For an interesting overview of the fields of research that could be important but are missed out see: 'Undone Science: Social Movements, Mobilized Publics, and Industrial Transitions', MIT Press Scholarship Online, Oxford Academic: https://academic.oup.com/mit-press-scholarship-online/book/22883

40 This also a pattern within higher education, which holds to an ideal of autonomy. It is common to hear complaints about managerialism, marketization and political interference, whether neoliberal in the West or nationalist in authoritarian countries. But, although such arguments often invoke society on their side, they rarely offer any evidence on what society actually thinks about ends and means. See for example: Thomas Docherty, *The New Treason of the Intellectuals: Can the University Survive?* (Manchester, UK: Manchester University Press, 2019).

41 Carolyn Marvin, *When Old Technologies Were New: Thinking About Electric Communication in the Late Nineteenth Century* (Oxford, UK: Oxford University Press, 1990).

42 Rebecca Masters et al., 'Return on Investment of Public Health Interventions: A Systematic Review', *Journal of Epidemiological Community Health* 71, no. 8 (1 August 2017): 827–34: https://doi.org/10.1136/jech-2016-208141
43 https://media.nesta.org.uk/documents/The_Biomedical_Bubble_v6.pdf
44 https://jech.bmj.com/content/71/8/827
45 https://www.health.org.uk/publications/the-need-for-a-complex-systems-model-of-evidence-for-public-health – which actually reflect the wider context in which people live.
46 N. Bloom, C.I. Jones, J. van Reenen and M. Webb, 'Are Ideas Getting Harder to Find Because of the Burden of Knowledge?', *American Economic Review* 110, no. 4.
47 Bloom et al., 'Are Ideas Getting Harder to Find?'
48 Benjamin F. Jones, 'The Burden of Knowledge and the "Death of the Renaissance Man": Is Innovation Getting Harder?', *Review of Economic Studies* 76, no. 1 (2009): 283–317; 'Age and Great Invention', *Review of Economics and Statistics*, MIT Press': https://direct.mit.edu/rest/article-abstract/92/1/1/57799/Age-and-Great-Invention?redirectedFrom=fulltext
49 Jan Brendel and Sascha Schweitzer, 'The Burden of Knowledge in Mathematics', *Open Economics* 2, no. 1 (1 January 2019): 139–49: https://doi.org/10.1515/openec-2019-0012; 'A Burden of Knowledge Creation in Academic Research: Evidence from Publication Data: Industry and Innovation', 28 (3): https://www.tandfonline.com/doi/abs/10.1080/13662716.2020.1716693
50 Kelsey Piper, 'Can a New Approach to Funding Scientific Research Unlock Innovation?', *Vox*, 18 December 2021: https://www.vox.com/future-perfect/2021/12/18/22838746/biomedicine-science-grants-arc-institute
51 Helen Buckley Woods and James Wilsdon, 'Experiments with Randomisation in Research Funding: Scoping and Workshop Report (RoRI Working Paper No.4)', report (Research on Research Institute, 18 April 2022): https://doi.org/10.6084/m9.figshare.16553067.v2
52 Ferric C. Fang, Anthony Bowen and Arturo Casadevall, 'NIH Peer Review Percentile Scores Are Poorly Predictive of Grant Productivity', *ELife* 5 (16 February 2016): e13323: https://doi.org/10.7554/eLife.13323
53 https://www.nber.org/books-and-chapters/economics-artificial-intelligence-agenda/impact-artificial-intelligence-innovation-exploratory-analysis
54 Fritz et al. (2019) provide good examples of how citizen data are currently being used around the world; for example, volunteers in the Philippines collecting household census data on poverty, nutrition, health, education, housing

and disaster risk reduction, which are then used by the Philippine Statistics Authority to enhance their statistics on 32 SDG indicators.

55 Alex Bell et al., 'Who Becomes an Inventor in America? The Importance of Exposure to Innovation': http://www.equality-of-opportunity.org/assets/documents/inventors_paper.pdf

10 The flawed reasoning of democracy

1 Foucault, *On the Government of the Living:* Lectures at the Collège de France, 1979–1980 (Basingstoke; New York: Palgrave Macmillan, 2014).

2 See for example: E. Schattschneider, *The Semi-Sovereign People: A Realist's View of Democracy in America* (New York: Holt, Rinehart, and Winston, 1960).

3 Yaron Ezrahi, 'Utopian and Pragmatic Rationalism: The Political Context of Scientific Advice', *Minerva* 18, no. 1 (1 March 1980): 111–31: https://doi.org/10.1007/BF01096662

4 'On Liberty', Econlib: https://www.econlib.org/library/Mill/mlLbty.html

5 The detailed story is in P. Jurs, 'Institution of Scientific Judgement', 2010.

6 'Views: Controlling Technology Democratically: The Paternalistic Doctrine of the Moral Responsibility of Science Has Led to a Chaotic Situation in Which Emotional Controversy Has Obscured Technical Information Vital to Democratic Control of Technology: https://www.jstor.org/stable/27845676

7 Claire Crawford, Lindsey Macmillan and Anna Vignoles, 'When and Why Do Initially High Attaining Poor Children Fall Behind?', n.d.

8 Olofsson et al., 'The Making of a Swedish Strategy'.

9 Robert A. Dahl, *On Democracy* (Newnhaven, CT: Yale University Press, 1998): https://doi.org/10.2307/j.ctv18zhcs4

10 Eric J. Gouvin, 'A Square Peg in a Vicious Circle: Stephen Breyer's Optimistic Prescription for the Regulatory Mess', *Harvard Journal on Legislation* 32 (n.d.).

11 Richard Bellamy, '"Dethroning Politics": Liberalism, Constitutionalism and Democracy in the Thought of F.A. Hayek', *British Journal of Political Science*, October 1994, 24(4): 419–41; Scott A. Boykin, 'Hayek on Spontaneous Order and Constitutional Design', *The Independent Review*, Summer 2010, 15(1):19–34.

12 Pierre Rosanvallon, *La contre-démocratie: La politique à l'âge de la défiance* (Paris: Seuil, 2006), English translation: *Counter-Democracy: Politics in an Age of Distrust* (New York: Cambridge University Press, 2008); Pierre Rosanvallon, *Le bon gouvernement* (Paris: Seuil, 2015).

13 Jon Elster, *Securities Against Misrule: Juries, Assemblies, Elections* (New York: Cambridge University Press, 2013).

14 Jon D. Miller, 'The Measurement of Civic Scientific Literacy', *Public Understanding of Science* (1998): Sci. 7 203DOI 10.1088/0963-6625/7/3/001

15 C. Funk, L. Rainie and D. Page, 'Public and Scientists' Views on Science and Society, *Pew Res. Center*, 1–111 (2015).

16 Misinformation is, of course, not new. See for example: Charlotte Sleigh, 'Fluoridation of Drinking Water in the UK, c. 1962–7. A Case Study in Scientific Misinformation Before Social Media', Royal Society (2022)

17 Massimiano Bucchi and Brian Trench, *Handbook of Public Communication of Science and Technology* (London: Routledge, 2 May 2008), pp. 111–30.

18 The combination of large language models like GPT, digital personal agents and sophisticated deliberation technologies such as Polis, point to a future where the trade-offs between mass participation and complexity are shifted.

19 Gerd Gigerenzer, Wolfgang Gaissmaier, Elke Kurz-Milcke, Lisa M. Schwartz and Steven Woloshin, 'Helping Doctors and Patients Make Sense of Health Statistics –2007': https://journals.sagepub.com/doi/full/10.1111/j.1539-6053.2008.00033.x; Ralph Hertwig, 'When to Consider Boosting: Some Rules for Policy-Makers', *Behavioural Public Policy* 1, no. 2 (November 2017): 143–61: https://doi.org/10.1017/bpp.2016.14

20 John Cook, Stephan Lewandowsky and Ullrich K. H. Ecker, 'Neutralizing Misinformation through Inoculation: Exposing Misleading Argumentation Techniques Reduces Their Influence', *PloS One* 12, no. 5 (2017): e0175799: https://doi.org/10.1371/journal.pone.0175799; Sander van der Linden, Jon Roozenbeek and Josh Compton, 'Inoculating Against Fake News About COVID-19', *Frontiers in Psychology* 11 (23 October 2020): 566790: https://doi.org/10.3389/fpsyg.2020.566790

21 David V. Budescu, Stephen Broomell and Han-Hui Por, 'Improving Communication of Uncertainty in the Reports of the Intergovernmental Panel on Climate Change': https://journals.sagepub.com/doi/10.1111/j.1467-9280.2009.02284.x

22 'Science Advice for Policy by European Academies', *Making Sense of Science for Policy Under Conditions of Complexity and Uncertainty*.

23 https://ourworldindata.org/faq-on-plastics#are-plastic-alternatives-better-for-the-environment

24 For a good overview see: S. Hecker, M. Haklay, A. Bowser, Z. Makuch, J. Vogel and A. Bonn, *Citizen Science: Innovation in Open Science, Society and*

Policy (London: UCL Press, 2018): https://doi.org/10.14324/111.978178735 2339

25 See: Oguz Acar, on 'Crowd Science and Science Skepticism', *Collective Intelligence*, Sage/ACM, 2023.

26 Though there are caveats if the topic is complex or it involves personal risk.

27 Political scientist Karen Stenner argued that authoritarian voters are best understood as 'simple-minded avoiders of complexity rather than closed-minded avoiders of change'.

28 A more activist view of democracy echoes the idea promoted by Pierre Rosanvallon that democracy is defined as much by the ways citizens can affect power as by the formal mechanisms; in other words: how they can survey and monitor; how they can resist; and how they can use the courts to bring politicians to judgement. Science and data play a part in all of these. Pierre Rosanvallon, *Counter-Democracy: Politics in an Age of Distrust*, trans. Arthur Goldhammer (New York: Cambridge University Press, 2008).

29 M. Rollwage, R.J. Dolan and S.M. Fleming, 'Metacognitive Failure as a Feature of Those Holding Radical Beliefs', *Current Biology* 28, no. 24: 4014–21.e8. doi: 10.1016/j.cub.2018.10.053. PMID: 30562522; PMCID: PMC6303190

30 Rollwage, Dolan and Fleming, 'Metacognitive Failure as a Feature of Those Holding Radical Beliefs'.

31 I have taught at each of these institutions and admire what they do. I wish we had an equivalent in the UK. Too many of our politicians become good at bluffing their way through situations rather than having deep understanding.

32 https://theippo.co.uk/net-zero-mobilising-knowledge-easier-effective-decision-making/

33 Zeynep Pamuk, *Science and Expertise* (Princeton, NJ: Princeton University Press, 2021).

34 Many countries now have policies and roadmaps for quantum, such as the US *National Strategic Overview for Quantum Information Science*. But I suspect that few legislators would be able to describe the implications.

35 Stefan P.L. de Jong, Jorrit Smit and Leonie van Drooge, 'Scientists' Response to Societal Impact Policies: A Policy Paradox, *Science and Public Policy* 43, Issue 1 (February 2016): 102–14: https://doi.org/10.1093/scipol/scv023

36 I later published an overview of what the science seemed to say, as a chapter in my book *Big Mind* and as a paper: https://www.nesta.org.uk/blog/meaningful-meetings-how-can-meetings-be-made-better/

11 The clash between global and national interest

1 Francisco J. Varela, *Ethical Know-How: Action, Wisdom, and Cognition* (Stanford, CA: Stanford University Press, 1992), p. 31. These lectures take further Varela's theory of 'enaction', the case for science closely attuned to lived experience, initially developed in: Francisco J. Varela, Evan Thompson and Elanor Rosch, *The Embodied Mind: Cognitive Science and the Human Experience* (Cambridge, MA: MIT Press, 1991).

2 James Poskett, *Horizons: The Global Origins of Modern Science* (Oxford, UK: Blackwells), p. 144.

3 It is a paradox that lumping together a huge diversity of different groups under the label 'indigenous' only really makes sense from the perspective of a dominant power. See for example: Sandra G. Harding, ed., *The Postcolonial Science and Technology Studies Reader* (Durham: Duke University Press).

4 Edgerton, 'From Innovation to Use', p. 126.

5 Reinhold Niebuhr, *Christian Realism and Political Problems* (New York: Charles Scribner's Sons, 1938), pp. 199, 556; Niehbuhr, 'The Tyranny of Science', *Theology Today* 10, no. 4 (January 1954): 464.

6 In his view, scientists were as likely to be producers of risks as producers of new answers – yet he also recognized that only scientists can then understand and contain the very risks that they produce, even as he worried that 'every bit of truth and wisdom must first be "cleared by science" before it can be given credence'.

7 See for example: https://www.ncbi.nlm.nih.gov/pmc/articles/PMC4644275/

8 Deborah Coen, *Climate in Motion, Science, Empire, and the Problem of Scale* (Chicago, IL: Chicago University Press, 2019).

9 A good recent overview is Simon Sharpe, *Five Times Faster: Rethinking the Science, Economics and Diplomacy of Climate Change* (Cambridge, UK: Cambridge University Press, 2023).

10 James Andrew Lewis, 'Technology and the Shifting Balance of Power', *Center for Strategic and International Studies (CSIS)*, 19 April 2022: https://www.csis.org/analysis/technology-and-shifting-balance-power; Daniel W. Drezner, 'Technological Change and International Relations', *International Relations* 33, no. 2 (2019): 286–303: https://journals.sagepub.com/doi/abs/10.1177/0047117819834629?journalCode=ireb.

11 'Survey of Chinese espionage activities in the United States since 2000', *Center for Strategic and International Studies*, Washington DC, 2023:

https://www.csis.org/programs/strategic-technologies-program/archives/survey-chinese-espionage-united-states-2000

12 Pierre-Bruno Ruffini, *Science and Diplomacy* (Cham: Springer International Publishing, 2017): https://doi.org/10.1007/978-3-319-55104-3

13 Treasury Board of Canada Secretariat, 'Experimentation Works', 24 June 2019: https://www.canada.ca/en/government/publicservice/modernizing/experimentation-works.html

12 Governing global science and technology

1 In a small way I've tried at various times in the past to develop ideas of this kind. When I ran the UK government's Strategy Unit under Tony Blair, we worked on reform options for the UN, and potential new arrangements for everything from nuclear proliferation and vaccines to organized crime. I've written chapters on global governance in books such as *Good and Bad Power*. More recently, I've attempted blueprints for specific areas of global governance (e.g. this one on global Internet governance), a more recent overview on options for governance of fields like cybersecurity and AI, and for reshaping parts of the UN system around data and knowledge.

2 https://changingchildhood.unicef.org/

3 'Why a Vaccine Hub for Low-Income Countries Must Succeed', *Nature* 607, no. 7918 (13 July 2022): 211–12: https://doi.org/10.1038/d41586-022-01895-6

4 Gregory D. Koblentz and Filippa Lentzos 'It's Time to Modernize the Bioweapons Convention', *Bulletin of the Atomic Scientists* (4 November 2016).

5 Maximilian Mayer, Mariana Carpes and Ruth Knoblich, *The Global Politics of Science and Technology: An Introduction* (Heidelberg: Springer, 2015).

6 As Robert Keohane pointed out in his book *After Hegemony*, the creation of new global institutions depends on strong leadership from a superpower, even if their operation does not. That leadership is missing right now. But, in all of these cases, there are grounds for confidence that national interests could be aligned with global ones.

7 See for example: L. Miao, D. Murray, W.S. Jung, V. Larivière, C.R. Sugimoto and Y.Y. Ahn, 'The Latent Structure of Global Scientific Development', *Nature Human Behaviour* 6, no. 9 (September 2022): 1206–17: doi: 10.1038/s41562-022-01367-x. Epub 2 June 2022. PMID: 35654964

8 These include the GPAI, PAI, AI Incidents Index, Responsible AI and many others. None have anything like the reach or legitimacy of the IPCC.

9 Such approaches might even be used to advance shared values, though that will be much harder: supporting human rights or avoiding retrograde actions such as closing the Internet.

10 Much of the academic work on international relations argues that science and technology hardly matter at all. Kalevi J. Holsti, for example, wrote that '(T)he foundational principles of international politics . . . did not undergo transformation as a result of major changes in the social and technological environment in previous centuries. Diplomatic life in 1775 was not unrecognizable from its predecessor in, let us say, 1700, despite that major intellectual upheaval of the eighteenth century, the Enlightenment. Similarly, the Industrial Revolution, surely a change as momentous as globalization is today, did not reorder major international institutions, except perhaps in the domain of war.' K.J. Holsti, *Taming the Sovereigns: Institutional Change in International Politics* (Cambridge, UK: Cambridge University Press, 2004), p. 19.

11 It would track and analyse global patterns and allow discussion of alternative possible orientations and portfolios for R&D in particular sectors and geographical contexts, working closely with the International Science Council, the International Network for Government Science Advice, OECD, UNESCO, as well as civil society, business, universities and other users of STI.

12 A recent article in *Nature* proposed one set of approaches: 'For example, at the global scale CGIAR and FAO among others could host a data repository for assessment and forecasting in coordination with IPCC, IPBES and others. Similarly, the HLPE (with support from FAO and other United Nations agencies) with an expanded mandate and additional resources can coordinate periodic assessment, forecasting, foresight and recommendations for policy actions in partnership with other SPSIs and the global research community (middle ring). The global-scale information can be fed by similar approaches at local and regional levels that will promote collaboration across all stakeholders to deliver functions for food transformation across all scales. FOLU, Food and Land Use Coalition; NGOs, non-governmental organizations; EFSA, European Food Safety Authority; EPA, United States Environmental Protection Agency; ReSAKKS, Regional Strategic Analysis and Knowledge Support System (for Africa); NZCILW, New Zealand Challenge Initiative on Land and Water; CCAP, China Centre for Agricultural Policy; FORAGRO, Forum for the Americas on Agricultural Research and Technology Development': https://www.nature.com/articles/s43016-022-00664-y

13 See 360 degree giving: https://www.threesixtygiving.org/

14 'Home – The Global Fund to Fight AIDS, Tuberculosis and Malaria': http://www.theglobalfund.org/en

15 These bodies are primarily accountable to their funders rather than to the public or potential beneficiaries and have been criticized for emphasizing the particular orientations for R&D favoured by these interests. But pooling resources can increase the impact of spending, and it is striking that it is missing in so many important areas – from gender equity to oceans – even though the sums involved are relatively small compared to overall R&D. There may be advantages in creating a menu of templates for such funds: providing model legal forms, model governance and decision-making structures, and protocols for the use of evidence and communication, for example. At present, each is bespoke, which means high transaction costs and unnecessary duplication.

16 'OECD Global Science Forum – OECD': https://www.oecd.org/sti/inno/global-science-forum.htm

17 'Global Observatory of Science, Technology and Innovation Policy Instruments (GO-SPIN)', UNESCO, 16 November 2016: https://en.unesco.org/go-spin

18 'NPO STS Forum': https://www.stsforum.org/

19 While there are many global gatherings around science and R&D, particularly academic gatherings such as the Society for Neuroscience (recent attendance of 30,000), European Society of Cardiology (32,000), and the American Chemical Society (15,000), there are no comparable meetings that connect to power, funding, policy and civil society, and none that look at R&D in the round.

20 One other important design principle follows from this. The most successful global institutions ensure a constant flow of personnel from and to national governments. The terms for their officials are limited – for example to five years in the case of the IAEA – which makes them very different to organizations like the WHO, more engaged and more responsive.

21 'Repertory of Practice of United Nations Organs – Codification Division Publications': https://legal.un.org/repertory/art108_109.shtml

13 Science, synthesis and metacognition

1 Many philosophers and theologians have criticized science and 'scientism' for these reasons, e.g. John Charles Knapp, 'Self-Deception and Moral Blindness in the Modern Corporation', who critiques four types of self-deception found in modern corporations: the self-deception of tribalism, the self-deception

of legalism, the self-deception of moral relativism and the self-deception of scientism.

2 Alex Marshall and Zoe Mou, 'Ibsen Play Is Canceled in China After Audience Criticizes Government', *New York Times*, 13 September 2018, sec. World: https://www.nytimes.com/2018/09/13/world/asia/china-ibsen-play.html

3 Christopher Ansell and Robert Geyer, '"Pragmatic Complexity": A New Foundation for Moving beyond "Evidence-Based Policy Making"?', *Policy Studies* 38, no. 2 (4 March 2017): 149–67: https://doi.org/10.1080/01442872.2016.1219033

4 Christl A. Donnelly et al., 'Four Principles to Make Evidence Synthesis More Useful for Policy', *Nature* 558, no. 7710 (June 2018): 361–4: https://doi.org/10.1038/d41586-018-05414-4

5 Peter D. Gluckman, Anne Bardsley and Matthias Kaiser, 'Brokerage at the Science–Policy Interface: From Conceptual Framework to Practical Guidance', *Humanities and Social Sciences Communications* 8, no. 1 (19 March 2021): 1–10: https://doi.org/10.1057/s41599-021-00756-3

6 My book, *The Art of Public Strategy*, sets out what this means at much greater length, and why understanding limitations of power and knowledge is so essential for understanding any issues of public policy.

7 There are huge numbers of options, with a sea of acronyms contributing to EBPM – MCA, CRELE, ACTA and others.

8 For a useful account of the difference between linear and iterative models, see the guidance note from the UNDESA Committee of Experts on Public Policy on Science Policy Interface (written by K. Allen) Strategy Note Science Policy Interface (March 2021).pdf (un.org)

9 D.J. Lang, A. Wiek, M. Bergmann, M. Stauffacher, P. Martens, P. Moll and C.J. Thomas, 'Transdisciplinary Research in Sustainability Science: Practice, Principles, and Challenges', *Sustainability Science* 7(S1) (2012): 25–43: https://doi.org/10.1007/s11625-011-0149-x

10 The first reactive synthesis competition (SYNTCOMP 2014) S. Jacobs, R. Bloem, R. Brenguier, R. Ehlers and T. Hell . . . *International Journal on Software Tools for Technology*, 2017.

11 There are many alternative methods for organizing meetings to help achieve synthesis: separating out roles (with someone to moderate, someone to be enthusiastic, another to be sceptical and so on) or to use antagonistic methods. These can be incorporated into formal bodies in the way that the UK Royal Commission on Environmental Pollution's (RCEP's) used 'interdisciplinary

deliberation' to encourage what Rein and Schon had called 'frame-reflection', the ability to reflect on how issues are framed, rather than treating this as a given.

12 See for example: on food systems: https://www.nature.com/articles/s43016-022-00664-y

13 NICE includes a formal process for engaging with the perspectives of different stakeholders to define the research questions and thus the evidence considered, how it is synthesized, and how it is interpreted. See: David Gough, 'Appraising Evidence Statements' (2021): https://doi.org/10.3102/0091732X20985072

14 I recently wrote a piece on some of the options for inquiries: https://theconversation.com/covid-crisis-what-kind-of-inquiry-do-we-need-to-learn-the-right-lessons-168163; IPPO also shared a broader overview of the issues and options: https://covidandsociety.com/what-we-want-covid-19-public-inquiries/

15 See Scott Page's book *Model Thinker* for a brilliant account of the virtues of using multiple models to grasp complex phenomena.

16 A good collection and overview is Dili Jeste et al., 'The New Science of Practical Wisdom', *Perspectives in Biology and Medicine* 62, no. 2: 216–36. Other useful overviews include: Stephen Hall, *Wisdom: From Philosophy to Neuroscience*, UQP 2010; The Arete Initiative at the University of Chicago, which called their $2 million research project into wisdom 'Defining Wisdom'; and Robert Sternberg, ed., *Wisdom: Its Nature, Origins and Development* (Cambridge, UK: Cambridge University Press, 1990), which provides over a dozen different definitions. There are also many widely used frameworks, such as Three-Dimensional Wisdom Scale; Berlin Wisdom Paradigm; the Balance Theory of Wisdom and many others, including the contribution of positive psychology in Christopher Peterson and Martin Seligman, *Character Strengths and Virtues*. Another source is https://evidencebasedwisdom.com/

17 'Wholesome Knowledge: Concepts of Wisdom in a Historical and Cross-Cultural Perspective': https://www.taylorfrancis.com/chapters/edit/10.4324/9781315789255-5/wholesome-knowledge-concepts-wisdom-historical-cross-cultural-perspective-aleida-assmann

18 This has been a common trope through the history of philosophy, including recent work on 'presence' (such as Derrida, and the movement of object-oriented ontology: see: Graham Harman, *Object Oriented Ontology* (Pelican Books, 2018).

19 This is a long-standing issue. A recent interesting paper looks at this in relation to genomics: https://www.matteotranchero.com/pdf/DataDrivenSearch_Tranchero_Mar2023.pdf

20 Scott E. Page, *The Model Thinker: What You Need to Know to Make Data Work for You* (Basic Books, 2021).

21 Page, *The Model Thinker.*

22 There is an extensive psychological literature on how, in some circumstances, suffering and trauma can aid psychological growth; see for example: E. Jayawickreme and L.E.R. Blackie, *Exploring the Psychological Benefits of Hardship: A Critical Reassessment of Posttraumatic Growth* (Switzerland: Springer, 2016).

23 See, for a not necessarily reliable source: https://www.cia.gov/readingroom/docs/0006542324.pdf

24 See for example, Andrew Targowski, *Cognitive Informatics and Wisdom Development: Interdisciplinary Approaches* (Hershey, PA and New York: Information Science Reference, 2011).

25 They might have been right about the direction of travel, but they tended to be wildly wrong on the speed of getting there.

26 Earlier versions of this section were published as a Nesta paper, and then as a chapter in my book *Social Innovation* (Bristol, UK: Policy, 2019).

27 Calestous Juma, *Innovation and Its Enemies: Why People Resist New Technologies* (Oxford, UK: Oxford University Press, 2016).

28 D. Edgerton, *The Shock of the Old* (Profile Books, 2019, updated edition), p. 6.

29 Such as the Nuffield Bioethics Council and the Ada Lovelace Institute for AI Ethics.

30 Leopold Aschenbrenner, 'Existential Risk and Growth': https://globalprioritiesinstitute.org/wp-content/uploads/Leopold-Aschenbrenner_Existential-risk-and-growth_.pdf

31 Rahul Bhatia, 'The Inside Story of Facebook's Biggest Setback', *Guardian*, 12 May 2016, sec. Technology: https://www.theguardian.com/technology/2016/may/12/facebook-free-basics-india-zuckerberg

Index